RESIDUE REVIEWS

VOLUME 53

RESIDUE REVIEWS

Residues of Pesticides and Other
Contaminants in the Total Environment

Editor
FRANCIS A. GUNTHER

Assistant Editor
JANE DAVIES GUNTHER

Riverside, California

VOLUME 53

SPRINGER-VERLAG
NEW YORK • HEIDELBERG • BERLIN
1974

Library of Congress Catalog Card Number 62-18595.
Printed in the United States of America.

New York: 175 Fifth Avenue, New York, N.Y. 10010
Heidelberg: 6900 Heidelberg 1, Postfach 1780, West Germany

ISBN 0-387-90084-5 Springer-Verlag New York Heidelberg Berlin
ISBN 3-540-90084-5 Springer-Verlag Berlin Heidelberg New York

---

Foreword

Worldwide concern in scientific, industrial, and governmental communities over traces of toxic chemicals in foodstuffs and in both abiotic and biotic environments has justified the present triumvirate of specialized publications in this field: comprehensive reviews, rapidly published progress reports, and archival documentations. These three publications are integrated and scheduled to provide in international communication the coherency essential for nonduplicative and current progress in a field as dynamic and complex as environmental contamination and toxicology. Until now there has been no journal or other publication series reserved exclusively for the diversified literature on "toxic" chemicals in our foods, our feeds, our geographical surroundings, our domestic animals, our wildlife, and ourselves. Around the world immense efforts and many talents have been mobilized to technical and other evaluations of natures, locales, magnitudes, fates, and toxicology of the persisting residues of these chemicals loosed upon the world. Among the sequelae of this broad new emphasis has been an inescapable need for an articulated set of authoritative publications where one could expect to find the latest important world literature produced by this emerging area of science together with documentation of pertinent ancillary legislation.

The research director and the legislative or administrative advisor do not have the time even to scan the large number of technical publications that might contain articles important to current responsibility; these individuals need the background provided by detailed reviews plus an assured awareness of newly developing information, all with minimum time for literature searching. Similarly, the scientist assigned or attracted to a new problem has the requirements of gleaning all literature pertinent to his task, publishing quickly new developments or important new experimental details to inform others of findings that might alter their own efforts, and eventually publishing all his supporting data and conclusions for archival purposes.

The end result of this concern over these chores and responsibilities and with uniform, encompassing, and timely publication outlets in the field of environmental contamination and toxicology is the Springer-Verlag (Heidelberg and New York) triumvirate:

Residue Reviews (vol. 1 in 1962) for basically detailed review articles concerned with any aspects of residues of pesticides and other chemical contaminants in the total environment, including toxicological considerations and consequences.

Bulletin of Environmental Contamination and Toxicology (vol. 1 in 1966) for rapid publication of short reports of significant advances and discoveries in the fields of air, soil, water, and food contamination and pollution as well as methodology and other disciplines concerned with the introduction, presence, and effects of toxicants in the total environment.

Archives of Environmental Contamination and Toxicology (vol. 1 in 1973) for important complete articles emphasizing and describing original experimental or theoretical research work pertaining to the scientific aspects of chemical contaminants in the environment.

Manuscripts for *Residue Reviews* and the *Archives* are in identical formats and are subject to review, by workers in the field, for adequacy and value; manuscripts for the *Bulletin* are not reviewed and are published by photo-offset to provide the latest results without delay. The individual editors of these three publications comprise the Joint Coordinating Board of Editors with referral within the Board of manuscripts submitted to one publication but deemed by major emphasis or length more suitable for one of the others.

June 26, 1974 Joint Coordinating Board of Editors

Preface

That residues of pesticide and other contaminants in the total environment are of concern to everyone everywhere is attested by the reception accorded previous volumes of "Residue Reviews" and by the gratifying enthusiasm, sincerity, and efforts shown by all the individuals from whom manuscripts have been solicited. Despite much propaganda to the contrary, there can never be any serious question that pest-control chemicals and food-additive chemicals are essential to adequate food production, manufacture, marketing, and storage, yet without continuing surveillance and intelligent control some of those that persist in our foodstuffs could at times conceivably endanger the public health. Ensuring safety-in-use of these many chemicals is a dynamic challenge, for established ones are continually being displaced by newly developed ones more acceptable to food technologists, pharmacologists, toxicologists, and changing pest-control requirements in progressive food-producing economies.

These matters are of genuine concern to increasing numbers of governmental agencies and legislative bodies around the world, for some of these chemicals have resulted in a few mishaps from improper use. Adequate safety-in-use evaluations of any of these chemicals persisting into our foodstuffs are not simple matters, and they incorporate the considered judgments of many individuals highly trained in a variety of complex biological, chemical, food technological, medical, pharmacological, and toxicological disciplines.

It is hoped that "Residue Reviews" will continue to serve as an integrating factor both in focusing attention upon those many residue matters requiring further attention and in collating for variously trained readers present knowledge in specific important areas of residue and related endeavors involved with other chemical contaminants in the total environment. The contents of this and previous volumes of "Residue Reviews" illustrate these objectives. Since manuscripts are published in the order in which they are received in final form, it may seem that some important aspects of residue analytical chemistry, biochemistry, human and animal medicine, legislation, pharmacology, physiology, regulation, and toxicology are being neglected; to the contrary, these apparent omissions are recognized, and some pertinent manuscripts are in preparation. However, the field is so large and the interests in it are so varied that the editors and the Advisory Board earnestly solicit suggestions of topics and authors to help make this international book-series even more useful and informative.

"Residue Reviews" attempts to provide concise, critical reviews of timely advances, philosophy, and significant areas of accomplished or needed endeavor in the total field of residues of these and other foreign chemicals in any segment of the environment. These reviews are either general or specific, but properly they may lie in the domains of analytical chemistry and its methodology, biochemistry, human and animal medicine, legislation, pharmacology, physiology, regulation, and toxicology; certain affairs in the realm of food technology concerned specifically with pesticide and other food-additive problems are also appropriate subject matter. The justification for the preparation of any review for this book-series is that it deals with some aspect of the many real problems arising from the presence of any "foreign" chemicals in our surroundings. Thus, manuscripts may encompass those matters, in any country, which are involved in allowing pesticide and other plant-protecting chemicals to be used safely in producing, storing, and shipping crops. Added plant or animal pest-control chemicals or their metabolites that may persist into meat and other edible animal products (milk and milk products, eggs, etc.) are also residues and are within this scope. The so-called food additives (substances deliberately added to foods for flavor, odor, appearance, etc., as well as those inadvertently added during manufacture, packaging, distribution, storage, etc.) are also considered suitable review material. In addition, contaminant chemicals added in any manner to air, water, soil or plant or animal life are within this purview and these objectives.

Manuscripts are normally contributed by invitation but suggested topics are welcome. Preliminary communication with the editors is necessary before volunteered reviews are submitted in manuscript form.

Department of Entomology
University of California
Riverside, California
June 26, 1974

F.A.G.
J.D.G.

Table of Contents

New aspects of organophosphorus pesticides. I. Introduction.

By

T. R. Fukuto*

In 1932, Willy Lange, an instructor in chemistry at the University of Berlin, and Gerda von Kreuger, a graduate student, described the synthesis of the organophosphorus esters dimethyl and diethyl phosphorofluoridate. Although this paper was concerned primarily with chemical synthesis, the following brief description of the physiological effects of these compounds was included: "*Interesting is the strong action of the dialkyl phosphorofluoridates on man. The vapor of these compounds had a pleasant aromatic odor. However, after a few minutes following inhalation, a strong pressure appears upon the larynx, combined with shortness of breath. Then cloudiness of vision and sensitivity of the eyes toward light occurs. After several hours these phenomena fade away. They did not appear to be caused by the acidic decomposition products of the ester but are probably attributable to the dialkyl phosphorofluoridate itself. The action was produced by a very small amount.*"

Evidently, this was the first account relating to the unusual physiological properties of specific types of organophosphorus compounds. The potential of organophosphorus esters as synthetic insecticides was recognized by Farbenfabriken Bayer, Germany, largely through the persistence of Lange, and in 1936 Gerhard Schrader began an intensive program on the synthesis and evaluation of organophosphorus esters as insecticides. Out of the early efforts, a number of effective insecticides were discovered including octamethyl pyrophosphoramide (schradan), tetraethyl pyrophosphate, and parathion.

Since Schrader's discoveries were revealed to the scientific community at the end of World War II, vast numbers of a variety of organophosphorus compounds have been synthesized and evaluated for insecticidal and other types of activity. During this period enormous progress has been made in the development of organophosphorus esters as insecticides and these materials represent one of the most important classes of organic insecticides in use today. In recent years

* Department of Entomology, University of California, Riverside, California 92502.

new organophosphorus compounds have been developed not only as effective insecticides, but also as fungicides, nematicides, herbicides, and other materials for agricultural use. Thus, organophosphorus compounds have proved to be unusually versatile materials which are currently being exploited for the control of a variety of pests.

The following seven articles in this volume originate from a symposium on "New Aspects of Organophosphorus Pesticides" held in Dallas, Texas at the 165th National Meeting of the American Chemical Society (Pesticide Division) in April, 1973. In the organization of this symposium, attempt was made to provide expertise in as many different facets of organophosphorus pesticides as possible with, of course, emphasis on recent advances. Two of the articles are concerned with the synthesis and insecticidal evaluation of a novel series of phosphoramidothioate derivatives (P. S. MAGEE) and phosphorus-containing imino oxazolidine and related compounds (E. J. GAUGHAN and J. J. MENN). In a somewhat different vein but still concerned with materials used for insect control are the articles on the metabolism of phosphonate insecticides (J. J. MENN and J. B. MCBAIN), oxidative rearrangement of phosphorothionate insecticides (D. A. WUSTNER, M. A. FAHMY, and T. R. FUKUTO), and organophosphorus insect chemosterilants (A. B. BORKOVEC). Turning to other pesticides, a comprehensive article on the metabolism and properties of the nematicide Nemacur (T. B. WAGGONER and A. M. KHASAWINAH) and an updated review on structure-activity relationships among organophosphorus fungicides (H. TOLKMITH and D. R. MUSSELL) are included. Although none of the articles are intended to be extensive reviews, they do, however, represent the cumulative efforts of the individual experts who participated in the symposium and in general contain the latest information on each subject.

Manuscript received September 4, 1973; accepted October 17, 1973.

New aspects of organophosphorus pesticides. II. Structure and bioactivity of Orthene® insecticide analogs.

By

PHILIP S. MAGEE*

Contents

I. Introduction

O,S-Dimethyl phosphoramidothioate was jointly commercialized in the United States as Monitor® Insecticide[1] by Chevron Chemical Company and Chemagro, and as Tamaron in Europe by Farbenfabriken Bayer. This exceptional insecticide/acaricide was discovered in 1964 by LORENZ at Bayer (SHRADER *et al.* 1972) and by MAGEE (1967 Patent-Monitor, 1972 Patent-Orthene) at Chevron Chemical. Both groups carried out extensive modifications within the class leading to subsequent patents. Based on simplicity, efficacy and economics, Monitor Insecticide survived as the most important member of its class and became a commercial reality in 1970. It has many virtues as a systemic broad-spectrum insecticide of good residual life and is expected to have a world-wide impact on many important crops. HAMMANN (1970) described the bioactivity of Monitor Insecticide in some detail. While safe to handle due to moderate dermal toxicity [LD_{50} (rats) 110 mg/kg (HAMMANN 1970)], Monitor and all of its simple analogs have relatively high oral toxicity [LD_{50} (rats) 20 mg/kg].

Discovery that *N*-acetylation greatly reduced mammalian toxicity

* Chevron Chemical Company, Richmond, California 94801.

[1] Monitor® Insecticide is a registered trademark of Chevron Chemical Company; use of the abbreviation "Monitor" refers to Monitor® Insecticide.

and, in some cases, did not lower insecticidal activity, was a major breakthrough. While the earliest members of the class lost activity in both areas, acetylation of Monitor Insecticide in 1968 led to the remarkable Orthene® Insecticide.[2] Orthene Insecticide is now being commercialized (1972) and possesses the best properties of Monitor coupled with exceptionally low mammalian toxicity [oral LD_{50} (rats) 900 mg/kg].

Like Monitor, the decision to commercialize Orthene was based on simplicity, efficacy, and economics. This decision was made after

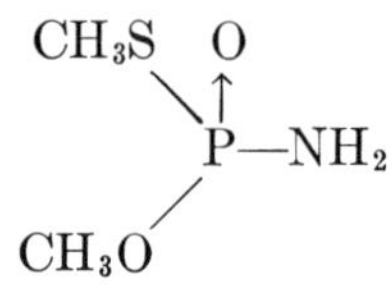

Monitor® Insecticide

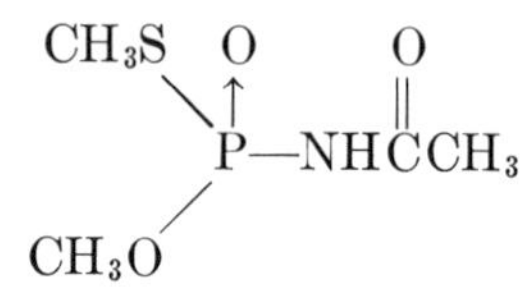

Orthene® Insecticide

a thorough study of the chemistry, physical properties, and biological activity of many Orthene analogs. The purpose of this report is to explore these variations in O, S, and N substitution.

Most of the desired structures in Table I were prepared by one of three methods: direct acylation of Monitor (A), acylation of Monitor precursor and either direct isomerization (B), or dealkylation-realkylation through an intermediate salt (C). As noted in Table II, yields ranged from one to 95 percent and were frequently below 20 percent. Fortunately, the unique polarity of these analogs relative

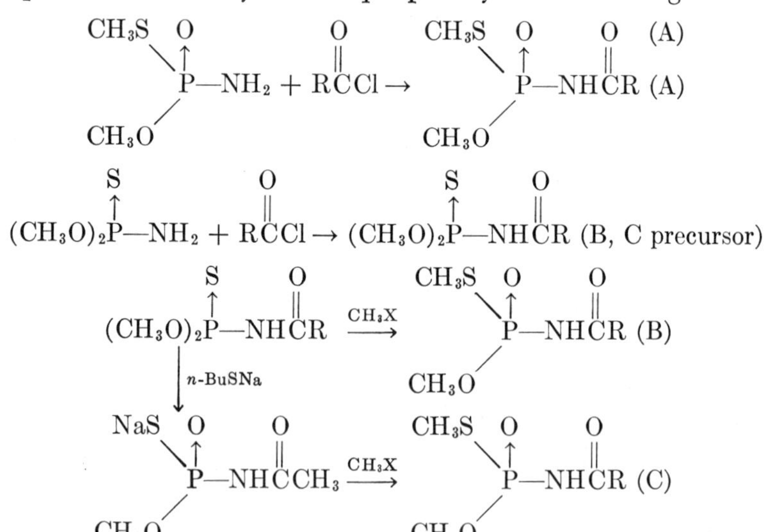

---

[2] Orthene® Insecticide is a registered trademark of Chevron Chemical Company for acephate; use of the abbreviation "Orthene" refers to Orthene® Insecticide.

to their precursors and by-products allowed clean separations in all cases. Even with liquid analogs rechromatography was rarely needed. By analysis, IR, NMR, and TLC purities of greater than 98 percent for all solids and greater than 95 percent for all liquids were established. Structures were unequivocal in all cases through corroboration of spectra (IR, NMR), physical properties (TLC, chromatography) and, in most cases, characteristic bioactivity. Note that the analytical data of Table II are recorded with one less significant number than usual. Phosphoramidates as a class are flame retardants and difficulty in combusting many of these compounds has reduced the accuracy of the data. In several cases, these data do not reflect the purity of the compounds as well as other methods (NMR, TLC).

A number of interesting side reactions were observed in the synthesis of these compounds. Some of the identified by-products reflect O- and S-attack, O,S-interchange, fragmentation, and radical involvement. Representative examples show some of the by-products observed:

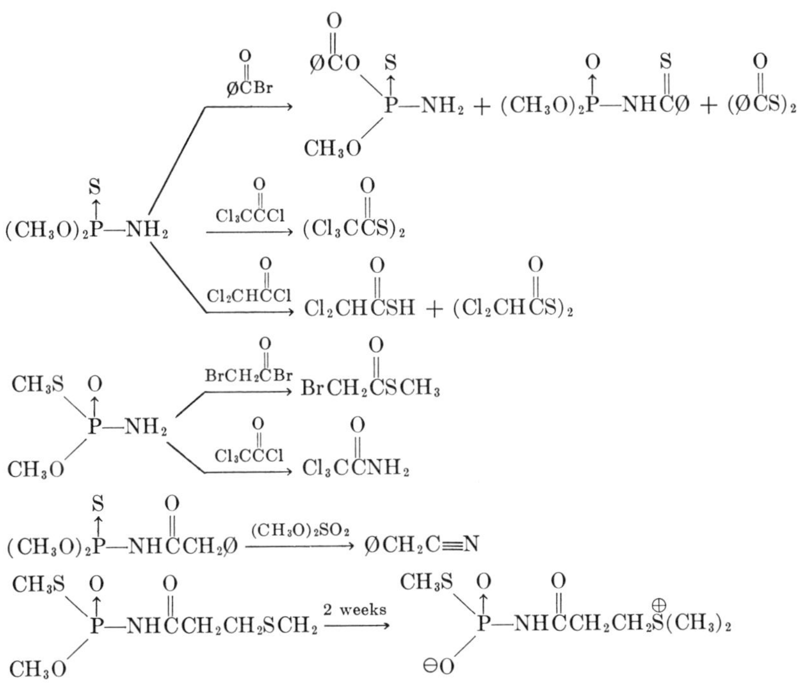

II. Experimental methods

a) *Synthesis*

The procedures given below cover most of the variations encountered in preparing the compounds of Table II. Acid chlorides were

Table I. *Insecticidal and mammalian toxicity of Orthene analogs.*

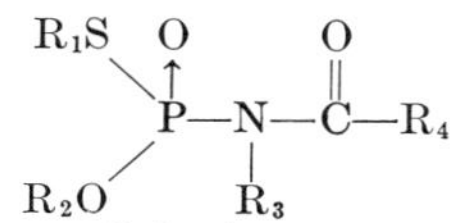

Compound	Substituents R_1	R_2	R_3	R_4	I/ED$_{50}$ ($\times 10^2$ ppm^{-1}) Houseflies[a]	2-Spotted mites[b]	Cotton aphid[c]	LD$_{50}$ (oral, rat[d]) (mg/kg)
1	CH_3	C_2H_5	H	CH_3	<0.2	<1	<3.3	500
2	C_2H_5	C_2H_5	H	CH_3	0.2	0.8	—	—
3	CH_3	CH_3	H	H	7.7	5.3	27	190
4	CH_3	CH_3	H	CH_3	20	50	250	900
5	CH_3	CH_3	CH_3	CH_3	<0.2	<3.3	<3.3	2,000
6	CH_3	CH_3	H	C_2H_5	7.2	40	170	>1,000
7	CH_3	CH_3	H	$(CH_2)_2CH_3$	8.3	23	67	125
8	CH_3	CH_3	H	$(CH_2)_4CH_3$	<1	34	15	25–60
9	CH_3	CH_3	H	$(CH_2)_6CH_3$	1	31	4	—
10	CH_3	CH_3	H	$(CH_2)_8CH_3$	1.1	16	21	125–250
11	CH_3	CH_3	H	$(CH_2)_9CH_3$	<1	100	220	—
12	CH_3	CH_3	H	$(CH_2)_{10}CH_3$	1.6	53	15	100–200
13	CH_3	CH_3	H	$(CH_2)_{11}CH_3$	<1	83	150	—
14	CH_3	CH_3	H	$(CH_2)_{12}CH_3$	1.1	91	100	100–250
15	CH_3	CH_3	H	$(CH_2)_{16}CH_3$	0.34	6.7	10	>250
16	CH_3	CH_3	H	$CH(CH_3)_2$	1.3	40	140	>500
17	CH_3	CH_3	H	$C(CH_3)_3$	0.2	1.3	17	—
18	CH_3	CH_3	H	$CH_2CH(CH_3)_2$	0.3	50	120	—
19	CH_3	CH_3	H	$CH(CH_2)_2CH_3$ $\vert$ CH_3	<0.2	1.3	<3.3	—

20	CH_3	CH_3	H	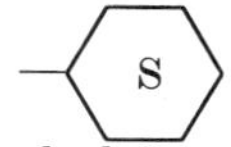	<0.2	1.1	2.5	—
21	CH_3	CH_3	H	oleyl[e]	1.4	67	5	—
22	CH_3	CH_3	H	$CH{=}C(CH_3)_2$	<0.2	<1	<3.3	—
23	CH_3	CH_3	H	$CH_2Ø$	<0.2	50	77	—
24	CH_3	CH_3	H	$CH_2CH_2Ø$	<0.2	1.1	<3.3	—
25	CH_3	CH_3	H	$CH{=}CHØ$	<0.2	12	<3.3	—
26	CH_3	CH_3	H	$CH_2SØ$	<0.2	6.7	5	—
27	CH_3	CH_3	H	$CH_2OØ$	3.0	56	2.5	—
28	CH_3	CH_3	H	CH_2Br	<0.2	<1	<3.3	—
29	CH_3	CH_3	H	CH_2Cl	17	16	36	—
30	CH_3	CH_3	H	$CHCl_2$	8.3	33	44	>500
31	CH_3	CH_3	H	$(CH_2)_2CH_2Cl$	4.5	10	500	3–4
32	CH_3	CH_3	H	$(CH_2)_9CH_2Br$	0.67	50	<3.3	—
33	CH_3	CH_3	H	CH_2OCH_3	22	50	220	200
34	CH_3	CH_3	H	$CH_2CH_2OCH_3$	1.7	1.3	<3.3	—
35	CH_3	CH_3	H	$CH_2CH_2OCH(CH_3)_2$	4.2	2.5	<3.3	—
36	CH_3	CH_3	H	$CH_2SC_2H_5$	2	16	1.7	—
37	CH_3	CH_3	H	$(CH_2)_3SCH_3$	0.2	0.7	2.5	—
38	CH_3	CH_3	H	OC_2H_5	6.7	<1	<3.3	—
39	CH_3	CH_3	H	$O(CH_2)_3CH_3$	0.3	<1	<3.3	—
40	CH_3	CH_3	H	SC_2H_5	16	8.3	29	—
41	$CH_2{=}CHCH_2$	CH_3	H	C_2H_5	8.3	40	4	125
42	$HC{\equiv}CCH_2$	CH_3	H	C_2H_5	10	26	220	500
43	$HC{\equiv}CCH_2$	C_2H_5	H	CH_3	1.4	0.7	—	—

[a] Microsprayed toxicant solution on anesthetized male and female flies; reading after 24 hours.
[b] Pinto bean leaves infested with mites were dipped in toxicant solution; reading after 24 hours.
[c] Cucumber leaves infested with aphids were dipped in toxicant solution; reading after 24 hours
[d] Sprague-Dawley rats, male and female.
[e] Oleyl: $—C{=}O(CH_2)_7CH{=}CH(CH_2)_7CH_3$.

Table II. *Physical and analytical data for Orthene analogs.*

Compound (see Table I)	Method	Yield (%)	m. p. (°C)	Phosphorus (%)		Sulfur (%)		Nitrogen (%)	
				Calculated	Found	Calculated	Found	Calculated	Found
1	B[a]	18	67–75	15.7	15.4	—	—	—	—
2	A, B[b]	4, *ca.* 50	Oil	14.7	14.6	—	—	6.6	6.5
3	B	30	Oil	—	—	—	—	8.3	8.1
4	A, B, C	50, 80, 95	90–91	16.9	16.7	17.5	17.8	—	—
5	A	34	Oil	15.7	14.8	16.3	17.5	—	—
6	B	15	38–42	15.7	15.3	16.3	15.7	—	—
7	A	15	Oil	—	—	14.8	14.8	6.6	5.8
8	A	44	Oil	—	—	13.4	12.5	5.9	5.3
9	A	28	Oil	11.6	10.5	—	—	5.3	5.3
10	B	50	Oil	10.5	10.7	10.9	10.4	—	—
11	A	20	58–61	10.0	9.8	—	—	4.5	4.3
12	B, C	45, 61	48–50	9.6	9.5	9.9	9.8	4.3	4.2
13	A	12	55–61	9.2	9.1	—	—	4.1	3.8
14	A	8	57–61	—	—	9.1	9.0	3.9	3.7
15	B	28	57–62	—	—	7.9	7.9	3.4	3.1
16	A	16	48–50	14.7	14.3	15.2	16.4	—	—
17	B	1	Oil	—	—	14.2	13.7	—	—
18	B	23	Oil	13.8	13.6	—	—	—	—
19	B	11	Oil	12.9	12.1	13.4	13.1	—	—
20	A	7	95–100	12.3	12.2	12.8	13.4	—	—
21	A	14	Oil	7.6	6.9	7.9	8.3	3.5	3.2
22	A	8	75–78	13.8	13.4	14.3	14.6	6.3	5.9
23	B	47	Oil	12.0	12.5	12.4	12.9	—	—
24	B	33	Oil	11.3	10.6	11.8	12.3	—	—

25	B	17	98–102	11.4	11.4	11.8	11.8	5.2	5.2
26	B	28	Oil	10.6	10.2	—	—	4.8	4.4
27	B	30	Oil	11.2	10.8	11.6	11.7	5.2	4.8
28	A	5	82–84	11.8	11.7	—	—	(% Br[d])	
29	A	2	113–114	14.0	14.1	(% Cl[e])		6.4	6.1
30	B	60	129–132	12.3	12.6	12.7	13.3	(% Cl[f])	
31	A	19	73–75	12.6	11.8	13.1	13.5	(% Cl[h])	
32	A	24	55–57	8.0	7.8	8.3	8.1	(% Br[g])	
33	B	29	48–50	14.5	14.8	15.0	14.9	6.6	6.3
34	B	39	Oil	13.6	13.2	—	—	6.2	5.9
35	A	39	Oil	12.1	11.9	12.6	12.1	—	—
36	A	19	Oil	12.7	11.7	—	—	5.8	5.2
37	A	50	76–78	12.0	12.3	—	—	5.4	5.1
38	B	50	Oil	14.6	14.0	15.0	15.1	—	—
39	B	37	Oil	12.8	12.5	—	—	5.8	5.7
40	B	14	Oil	13.5	12.8	—	—	6.1	6.2
41	C	74	Oil	—	—	14.4	14.1	6.3	6.3
42	C	70	semi-solid	—	—	14.5	14.2	6.3	6.1
43	B[c]	*ca.* 80	Oil	14.1	13.8	14.9	15.6	—	—

[a,b,c] Isomerized by heating with excess CH_3I^a, $C_2H_5I^b$, and $HC{\equiv}CH_2Br^c$.
[d] Bromine: calc. 30.2, found 28.1 %.
[e] Chlorine: calc. 16.3, found 16.5 %.
[f] Chlorine: calc. 28.1, found 27.2 %.
[g] Bromine: calc. 20.6, found 21.8 %.
[h] Chlorine: calc. 14.4, found 14.4 %.

used as purchased or prepared from the purchased acids (Nos. 22, 32). Except for methoxyacetyl chloride (Aldrich Chemical Co.), all of the O or S hetero acid chlorides were prepared by nucleophilic displacements on halo-acids, esters, or nitriles or by conjugate addition to acrylic esters (Nos. 26, 27, 34 to 37). These acid derivatives are well-known and the methods are readily available. Elemental analyses were difficult, especially in some of the smaller members, as phosphoramidates are flame retardants. Special methods were developed during this project to improve the accuracy of the figures. Independent of the analyses, however, structure and purity were unequivocally determined by NMR and IR spectra. The characteristic CH_3OP and CH_3SP doublets gave good integrations with the acyl group (Varian T-60 NMR), while the positions of the NH, amide C=O, P→O, and POC frequencies were nearly identical in each case (Perkin-Elmer 137 IR). In all examples where chromatography was used (gas, column, or TLC), retention behavior was characteristic. Column chromatography was exceptionally effective in separating analogs in >95 percent purity even when the yield was below five percent as none of the starting materials or by-products have similar retentions. Care should be exercised in handling Monitor Insecticide (oral LD_{50} = 20 mg/kg) and the chlorobutyryl analog of Orthene Insecticide (oral LD_{50} = 3 to 4 mg/kg).

b) Intermediates

1. ***O,O*-Dimethyl decanoylphosphoramidothioate.** — An equimolar mixture of *O,O*-dimethyl phosphoramidothioate and *n*-decanoyl chloride was refluxed in methylene chloride for two hours. After stirring with water and extracting with methylene chloride, the organic phase was washed with water, dried over magnesium sulfate, and stripped. The pale yellow oil (96 percent) solidified on standing to a waxy solid, m.p. 25° to 30°C. Calc. for $C_{12}H_{26}NO_3PS$: N 4.80, P 10.50, S 10.85; found: N 4.82, P 10.32, S 11.20.

2. ***O,O*-Dimethyl formylphosphoramidothioate.** — Ten ml of formic acid was added to 20 ml of acetic anhydride cooled in ice. The mixture was heated to 50°C/15 min. to generate the mixed anhydride, then recooled in ice. Thirty ml of methylene chloride, 28.2 g (0.20 mole), of *O,O*-dimethyl phosphoramidothioate, and 0.5 ml of 85 percent phosphoric acid were added and the system was allowed to warm to room temperature. Reaction was complete within 23 hours by glc analysis. The mixture was cooled and added to 15 g of ice plus 30 ml of saturated NH_4Cl, followed by addition of 45 ml of 15 percent NH_4OH, with cooling, to neutralize. The organic phase was separated and the aqueous phase was extracted once with methylene chloride. One washing with 30 ml of saturated NH_4Cl

was followed by drying ($MgSO_4$) and stripping. The clear yellow oil, 25.0 g (98 percent), contained 95 percent *N*-formyl and five percent *N*-acetyl derivative by glc assay. At reflux, the reaction was complete in one hour, but the product contained 20 percent *O,O*-dimethyl acetylphosphoramidothioate.

3. *O*-Methyl *S*-sodio acetylphosphoramidothioate. — A slight excess of *n*-propyl or *n*-butyl mercaptan was added to a cooled equimolar solution of *O,O*-dimethyl acetylphosphoramidothioate and 50 percent NaOH in methanol. After refluxing for three hours and standing overnight to allow complete salt separation, the solid was removed, washed with cold methanol, and dried in a desiccator. The first crop yield (60 to 90 percent) varied with concentration as the salt is somewhat soluble in methanol. Calc. for $C_3H_7NNaO_3PS$: S 16.75, P 16.20; found: S 16.75, P 15.90. In D_2O, the NMR spectrum was definitive, showing a CH_3OP doublet near 3.7 ppm ($J_{P\text{-}H} = 13$ cps), a $CH_3(C{=}O)NH{-}P$ doublet near 2.2 ppm ($J_{P\text{-}H} = 1\text{–}2$ cps), and HDO from NH exchange. A Nujol mull showed NH (3300 cm^{-1}), a C=O doublet (1650 cm^{-1}), a P→O doublet (1150 cm^{-1}), and $P{-}OCH_3$ (*ca.* 1030 cm^{-1}).

c) Orthene® Insecticide analogs

1. *O,S*-Dimethyl 11-bromoundecanoylphosphoramidothioate (*A*, No. 32). — An equimolar mixture of *O,S*-dimethyl phosphoramidothioate and 11-bromoundecanoyl chloride was refluxed in methylene chloride for three hours. Addition of magnesium sulfate to the mixture assured anhydrous conditions and often improved the yield. The solution was stirred with water and neutralized with sodium carbonate, resulting in a serious emulsion. Addition of large amounts of magnesium sulfate was necessary to complex the aqueous phase and allow a separation. After drying and stripping, the crude oil was chromatographed through a Merck (Darmstadt) silica column (10 × 1.5 inches). After removal of less polar by-products with methylene chloride, the product was eluted with 25 percent acetone in methylene chloride as a viscous oil which crystallized on standing; yield 24 percent, m.p. 55° to 57°C. Calc. for $C_{13}H_{27}BrNO_3PS$: S 8.25, P 7.96, Br 20.58; found: S 8.08, P 7.77, Br 21.75.

2. *O,S*-Dimethyl 4-methylthiobutyrylphosphoramidothioate (*A*, No. 37). — *O,S*-Dimethyl phosphoramidothioate and a 25 percent excess of 4-methylthiobutyryl chloride were refluxed in methylene chloride for three hours. The solution was added to water, neutralized with sodium carbonate, and the aqueous phase was extracted with more methylene chloride. Drying over magnesium sulfate and stripping gave the crude product as a yellow oil. This was chromatographed as described in the example above, giving a viscous oil which crystallized on standing; yield 50 percent, m.p. 76° to 78°C. Calc. for $C_7H_{16}NO_3PS_2$: N 5.45, P 12.02; found: N 5.12, P 12.27.

3. *O,S*-Dimethyl dodecanoylphosphoramidothioate (*B*, No. 12).— Sixteen g of *O,O*-dimethyl dodecanoylphosphoramidothioate was combined with 4.0 g of dimethyl sulfate in ten ml of chloroform and refluxed for three hours in a hot water bath (80° to 85°C). Reaction progress was readily followed by NMR as development of the CH_3SP doublet, by IR as development of a P → O bond, or by TLC. The product solution was chromatographed through a 10 × 1.5 inch column of Merck (Darmstadt) silica using a graded series of solvents. As eluted in 45 percent yield with 25 percent acetone in methylene chloride, the phosphoramidothioate was contaminated with about ten percent lauric acid. This problem could have been avoided by carbonate treatment of the crude product solution. However, a second chromatography was used in this case to provide a purified product, m.p. 45° to 47°C. Calc. for $C_{14}H_{30}NO_3PS$: P 9.56, S 9.92; found: P 9.80, S 9.93.

4. *O,S*-Dimethyl dichloroacetylphosphoramidothioate (*B*, No. 30).—Nine g of *O,O*-dimethyl dichloroacetylphosphoramidothioate was mixed with 3.0 g of dimethyl sulfate, heated to 80°C briefly to melt the solid, then cooled to 60°C for the reaction duration. External heat was provided by an oil bath while an internal cooling coil coupled to a thermoregulated pump was used to remove exothermicity. In this case, there was little exotherm, but in others this type of control can prevent a run-away decomposition. At one hour 67 percent isomerization was observed by NMR integration of the ratio $CH_3SP/CHCl_2$. On cooling, the product mixture was largely solid. This was dissolved in chloroform, treated with carbon and crystallized by reducing the volume and adding hexane. The product separated in satisfactory purity, yield 60 percent, m.p. 129° to 132°C. Calc. for $C_4H_8Cl_2NO_3PS$: S 12.71, Cl 28.10, P 12.28; found: S 13.27, Cl 27.20, P 12.60. Both the NMR spectrum and the IR spectrum (mull) were exceptionally sharp and definitive.

5. *O*-Methyl *S*-allyl propionylphosphoramidothioate (*C*, No. 41).— Five g of *O*-methyl S-sodio propionylphosphoramidothioate was refluxed in 30 ml of allyl bromide for three hours. After cooling, the sodium bromide was removed by filtration and the product stripped at 60°C/1 mm. Filtration through Celite to remove a small amount of NaBr gave the nearly pure product as a clear brown oil; yield 4.0 g (74 percent). Further purification was readily accomplished by chromatography. Calc. for $C_7H_{14}NO_3PS$: C 37.67, H 6.38, N 6.28, S 14.38; found: C 37.20, H 6.15, N 6.28, S 14.05.

6. *O,S*-Dimethyl dodecanoylphosphoramidothioate (*C*, No. 12).— An equimolar mixture (0.062 mole) of *n*-propyl mercaptan, 50 percent NaOH, and *O,O*-dimethyl dodecanoylphosphoramidothioate in 100 ml of methanol was refluxed for three hours and allowed to stand overnight. No attempt is made to isolate the S-sodio salt, which has detergent-like qualities and is physically difficult to handle. Dimethyl sulfate (15.6 g, 0.124 mole) was added (exotherm to *ca.* 35°C) and the mixture was stirred at room temperature for 1.5 hours. Dilution with 50 ml of water and partial stripping of the methanol was followed by extraction with methylene chloride. The extract was washed with water, dried, and stripped to a viscous oil (21.5 g) which gave a butter-like solid on standing. Pure material was obtained by crystallization from hexane, yield 61 percent, m.p. 49° to 50°C. Calc. for $C_{14}H_{30}NO_3PS$: P 9.56, S 9.92; found: P 9.57, S 9.97.

III. Results and discussion

One possible explanation for the bio-activity of Orthene Insecticide and its analogs is simply that of degradation to Monitor Insecticide. Unpublished work in our laboratories by CROSSLEY (1973) on the hydrolytic breakdown of Orthene shows no formation of Monitor at any pH. However, when topically applied to pinto bean leaves, there are varying degrees of biochemical breakdown to Monitor depending on the Orthene structure. Thus, the dichloroacetyl analog (No. 30) generated 24.9 percent Monitor after nine days, while Orthene formed only 3.7 to 4.8 percent. A few generated less Monitor, but the majority of 18 analogs tested gave values between these two levels. As two test species (mites and aphids) were evaluated by feeding on treated pinto bean leaves, it was of interest to compare the generated ED_{50}'s with the amount of biochemically formed Monitor. There was no discernible correlation. In fact, some of the most active analogs corresponded to near-zero formation of Monitor Insecticide in the bean leaf. Thus, the activity of Orthene analogs is clearly due to their total structure and not the amount of Monitor Insecticide formed prior to ingestion by the pest.

Discussion of bioactivity of Orthene analogs is on a logical structural basis, referring to selected compounds from Table I. Thus, the effects of *O,S,N*-substitution, acyl chain length, acyl branching and cyclization, acyl unsaturation, and hetero-atoms and halogens in the acyl group, will be considered separately.

Original ED_{50} data have been converted to relative toxicities by simply inverting the scale, that is, relative toxicity = $10^2/ED_{50}$ ppm^{-1}. This is visually easier to handle as the highest activities are associated with the largest numbers. For a quick evaluation of any member of Table I, comparison with Orthene Insecticide (No. 4) is useful. In its major uses, Orthene Insecticide is effective on crops at 0.5 to 1.0 lb/acre. Therefore, a compound with 25 to 50 percent of Orthene activity still falls in the range of present-day commercial insecticides.

In varying substitution on *O* and S, it was quickly apparent that activity maximized at *O,S*-dimethyl. In fact, the drop in activity from

O,S-dimethyl to O,S-diethyl is extraordinary when compared to the same changes in Monitor Insecticide:

Orthene	OCH_3/SCH_3 (No. 4)	OC_2H_5/SCH_3 (No. 1)	OC_2H_5/SC_2H_5 (No. 2)
F/M/A[3]	20/50/250	<0.2/<1/<3.3	0.2/0.8/—
Monitor	OCH_3/SCH_3	OC_2H_5/SCH_3	OC_2H_5/SC_2H_5
F/M/A[3]	56/32/53	39/—/20	25/—/10

The effect was much less when CH_3S is replaced by an unsaturated group such as allylthio (No. 41) or propargylthio (No. 42). These examples have N-propionyl groups and must be compared with the corresponding O,S-dimethyl analog (No. 6). It is readily seen that while aphicidal activity dropped sharply with allylthio, it remained about the same with propargylthio. Considering the extreme sensitivity to the change $CH_3S \rightarrow C_2H_5S$, it is remarkable that the changes $CH_3S \rightarrow HC{\equiv}CCH_2S$ and $H_2C{=}CHCH_2S$ had so little effect. Of further interest was the great reduction of activity in the propargylthio analog on changing $CH_3O \rightarrow C_2H_5O$ (No. 43), again showing the extreme sensitivity to changes from methyl to ethyl. Replacement of NH by NCH_3 (No. 5) reduced the insecticidal activity to a low level and the mammalian toxicity from 900 to 2,000 mg/kg. The effect on Monitor Insecticide was similar.

Recognizing that activity maximizes at O,S-dimethyl, the remaining studies involve systematic changes in the acyl group. Replacement of N-acetyl by N-formyl (No. 3) caused a major reduction of insecticidal activity while mammalian toxicity increased four-fold. In the other direction, increasing acyl chain length was studied in compounds No's. 6 to 15 (N-propionyl to N-octadecanoyl). While generally decreasing activity, with some specific exceptions, the effect of the change $NHC(O)CH_3 \rightarrow NHC(O)R_4$ ($R_4{=}C_2H_5{-}C_{17}H_{35}$) was far less than that of $CH_3X \rightarrow C_2H_5X$ (X=O or S)! This remarkable set of facts coupled by maximum activity in unbranched chains almost suggests a biological arrow with a phosphoramidate head and very critical dimensions. In scanning the effects of increasing acyl chain length, it is clear that (1) activity against flies dropped rapidly to a low value ranging from $R_4{=}C_5$ to C_{17}, (2) miticidal activity remained high with several members better than Orthene Insecticide out to $R_4{=}C_{13}$ with still significant activity at $R_4{=}C_{17}$ (No. 15) but with the drop obvious, (3) aphicidal activity was variable from very low ($R_4{=}C_7$, No. 9) to very high ($R_4{=}C_{10}$, No. 11) with some evidence of an odd-even effect in the range $R_4{=}C_9$ to C_{13} (Nos. 10 to 14), and mammalian toxicity increased discontinuously between $R_4{=}C_2$ and C_3, with toxicity maximizing near $R_4{=}C_5$ and only dropping again above $R_4{=}C_{13}$.

Branching and cyclization (a form of branching) in the acyl group proved interesting. The following comparisons are instructive:

[3] F/M/A is an abbreviation for Flies/Mites/Aphids.

$\overset{O}{\overset{\|}{C}}R$	$\overset{O}{\overset{\|}{C}}CH_2CH_3$ (No. 6)	$\overset{O}{\overset{\|}{C}}CH(CH_3)_2$ (No. 16)	$\overset{O}{\overset{\|}{C}}C(CH_3)_3$ (No. 17)
F/M/A[3]	7.2/40/170	1.3/40/140	0.2/1.3/17

$\overset{O}{\overset{\|}{C}}R$	$\overset{O}{\overset{\|}{C}}CH_2CH(CH_3)_2$ (No. 18)	$\overset{O}{\overset{\|}{C}}CH(CH_2)_2CH_3$ with CH_3 branch (No. 19)	$\overset{O}{\overset{\|}{C}}$— 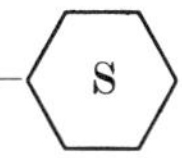 (No. 20)
F/M/A[3]	0.3/50/120	<0.2/1.3/<3.3	<0.2/1.1/2.5

A single terminal branch on a short acyl chain (Nos. 16 and 18) caused significant reduction in toxicity to flies. Double branching (No. 17), internal branching on a longer acyl (No. 19), or cyclization (No. 20) led to greatly reduced insecticidal activity. In the light of observed high activity out to very long acyl chain lengths, the effect on activity by a single *alpha*-methyl group is remarkable. Again, exceptional steric requirements for activity are implied.

Unsaturation in the acyl groups produced varying effects depending on location. Thus, conjugated unsaturation (No. 22) nearly destroyed activity, while an olefinic bond near the middle of the chain (No. 21) simply modified it:

$\overset{O}{\overset{\|}{C}}R$	$\overset{O}{\overset{\|}{C}}CH_2CH(CH_3)_2$ (No. 18)	$\overset{O}{\overset{\|}{C}}CH{=}C(CH_3)_2$ (No. 22)
F/M/A[3]	0.3/50/120	<0.2/<1/<3.3

$\overset{O}{\overset{\|}{C}}R$	$\overset{O}{\overset{\|}{C}}(CH_2)_{16}CH_3$ (No. 15)	$\overset{O}{\overset{\|}{C}}(CH_2)_7CH{=}CH(CH_2)_7CH_3$ (No. 21)
F/M/A[3]	0.34/6.7/10	1.4/67/5

Unsaturation due to phenyl groups reduced bio-activity, the effect depending strongly on the position. Thus, in the *N*-phenylacetyl analog (No. 23), activity on flies was greatly reduced while miticidal and aphicidal activity remained high. In the propionyl analog, the *beta*-phenyl group (No 24)

$\overset{O}{\overset{\|}{C}}R$	$\overset{O}{\overset{\|}{C}}CH_2\emptyset$ (No. 23)	$\overset{O}{\overset{\|}{C}}CH_2CH_2\emptyset$ (No. 24)	$\overset{O}{\overset{\|}{C}}CH{=}CH\emptyset$ (No. 25)
F/M/A[3]	<0.2/50/77	<0.2/1.1/<3.3	<0.2/12/<3.3

caused more general suppression of activity. Introduction of conjugated unsaturation led to a selective increase in miticidal activity, a surprising result in view of its effect in the 3,3-dimethylacryloyl analog (No. 22). All of these effects with double bonds and phenyl groups were too variable to allow predictability in new synthetic analogs. No generalizations are possible.

While investigating the effect of phenyl substituents, oxygen and sulfur were substituted for methylene in hopes of observing potentiation of activity. The effect was sizable, especially in augmenting acaricidal activity:

$\overset{O}{\parallel}$ CR	$\overset{O}{\parallel}$ $CCH_2CH_2Ø$ (No. 24)	$\overset{O}{\parallel}$ $CCH_2SØ$ (No. 26)	$\overset{O}{\parallel}$ $CCH_2OØ$ (No. 27)
F/M/A[3]	<0.2/1.1/<3.3	<0.2/6.7/5	3.0/56/2.5

The miticidal activity of the *N*-phenoxyacetyl analog is equal to that of Orthene, a 50-fold increase over the *N*-3-phenylpropionyl analog. These observations led to a more detailed search for a heteroatom effect in the purely aliphatic analogs where bio-activity is expected to be higher. Both sulfur and oxygen showed a positive enhancement with the heteroatom in the *alpha*-position. While the effect with sulfur was moderate, that with oxygen was extraordinary. Greatly suppressing activity in the *beta*-position (Nos. 34 and 35), *alpha*-oxygen in the *N*-methoxyacetyl analog (No. 33) generated activity equal to Orthene Insecticide and much higher than its nearest saturated analog (compare with *N*-butyryl. No. 7):

$\overset{O}{\parallel}$ CR	$\overset{O}{\parallel}$ CCH_2OCH_3 (No. 33)	$\overset{O}{\parallel}$ $CCH_2CH_2OCH_3$ (No. 34)
F/M/A[3]	22/50/220	1.7/1.3/<3.3

O,S-Dimethyl methoxyacetylphosphoramidothioate is the most outstanding broad-spectrum insecticide discovered in the analog study. While equal to Orthene Insecticide in bio-activity and with excellent residual life, it is more expensive to produce and somewhat more toxic.

Halogens in the acyl group were expected to produce sizable effects and some were indeed found, though the observations are restricted to a few bromine and chlorine analogs. Both bromine and chlorine reduced insecticidal activity in the simple acetyl analogs:

$\overset{O}{\parallel}$ CR	$\overset{O}{\parallel}$ CCH_2Br (No. 28)	$\overset{O}{\parallel}$ CCH_2Cl (No. 29)	$\overset{O}{\parallel}$ $CCHCl_2$ (No. 30)
F/M/A[3]	<0.2/<1/<3.3	17/36/36	8.3/33/44

$\overset{O}{\overset{\|}{C}}R$	$\overset{O}{\overset{\|}{C}}(CH_2)_9CH_3$ (No. 11)	$\overset{O}{\overset{\|}{C}}(CH_2)_9CH_2Br$ (No. 32)
F/M/A[3]	<1/100/220	0.67/50/<3.3
$\overset{O}{\overset{\|}{C}}R$	$\overset{O}{\overset{\|}{C}}CH_2CH_2CH_3$ (No. 7)	$\overset{O}{\overset{\|}{C}}CH_2CH_2CH_2Cl$ (No. 31)
F/M/A[3]	8.3/23/67	4.5/10/500
Oral rat	125	3 to 4

In the case of bromine, hydrolytic degradation prior to ingestion could account for the activity loss in the mite and aphid tests. However, flies were tested by direct microspray and degradation alone cannot account for the 100-fold drop in activity here. In a longer acyl chain, bromine was again found to reduce activity compared to its unsubstituted relative, one of the best insecticides found in this study. In a single example, chlorine in the *N*-4-chlorobutyryl analog gave remarkable potentiation of aphicidal activity and mammalian toxicity. Relative to its unchlorinated analog, aphicidal activity increased over sevenfold and was twice as active as Orthene Insecticide. The toxicity of this analog is hard to understand. Our first value of <5 mg/kg was carefully verified on a highly purified sample (chromatographed and recrystallized). This level of toxicity is extraordinary in a class where most members display values from 100 to 1,000 mg/kg. Even the parent Monitor Insecticide class is generally in the range of five to 25 mg/kg. More work in this immediate area, e.g., 3-chloropropionyl, 5-chlorovaleryl, and others will be needed to determine the uniqueness of this structure.

Carbamate and thiolcarbamate analogs of Orthene proved disappointing as can be seen by reviewing Nos. 38–40. While the thiolcarbamates were more active than the carbamates, only the S-ethyl analog (No. 40) had significant activity. Higher homologs (not shown in Table I) were found to be weak or inactive. The series was especially disappointing in view of the reported usefulness of the all-oxygen analogs, K-20-35 and Avenin as systemic insecticides of low mammalian toxicity (DERKATSCH 1969, p. 428):

$$(CH_3O)_2\overset{O}{\overset{\uparrow}{P}}\text{—}NH\overset{O}{\overset{\|}{C}}OCH_3 \qquad [(CH_3)_2CHO]_2\overset{O}{\overset{\uparrow}{P}}\text{—}NH\overset{O}{\overset{\|}{C}}OCH_3$$

K-20-35 Avenin

IV. Conclusions

This study of Orthene Insecticide analogs has developed some remarkable observations on structural limitations for high activity within the class. It has also uncovered several analogs of outstanding specific and general activities:

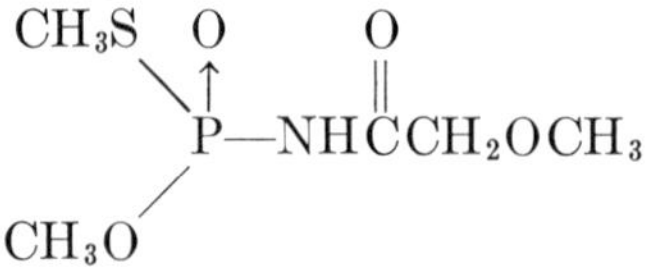

broad-spectrum insecticide/acaracide

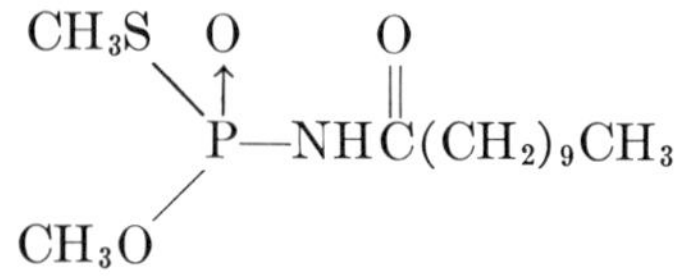

acaracide/aphicide

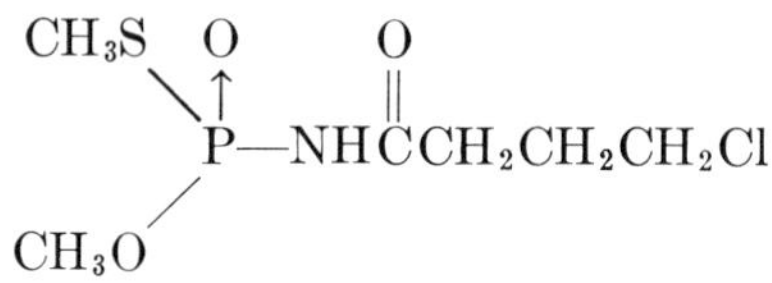

aphicide

Acknowledgment

The author wishes to thank Mrs. Zenaida dela Paz for her excellent technical assistance.

Summary

Analogs of Orthene® Insecticide (*O,S*-dimethyl acetylphosphoramidothioate) with variations of *O*, *S*, and *N* substitution were prepared and tested on several insects. Highest activity was associated with *O,S*-dimethyl analogs having relatively simple, saturated linear *N*-acyl groups. In fact, activity remained high to chain lengths $>C_{14}$, some members being better miticides than Orthene Insecticide and nearly equivalent aphicides. *Alpha*-methoxy and *gamma*-chloro substitutions gave remarkable enhancements while branching, cyclization, conjugated unsaturation, phenyl groups, *beta*-alkoxy, and *alpha*-halo groups generally reduced bioactivity. With few exceptions, the best analogs of Orthene Insecticide were found to combine high insecticidal activity with moderate to low mammalian toxicity.

References

CROSSLEY, J.: Unpublished data (1973).

DERKATSCH, G. I.: Phosphorus isocyanates. Angew. Chem. Internat. Ed. **8**, 421 (1969).

HAMMANN, I.: Tamaron, a new insecticide and acaricide. Pflanzenschutz-Nachrichten "Bayer" **23**, 133 (1970).

MAGEE, P. S.: U.S. Pat. 3, 309, 266 (Monitor) (1967).

—— U.S. Pat. 3, 649, 723 (Orthene) (1972).

SCHRADER, G., W. LORENZ, G. UNTERSTENHOFER, and I. HAMMANN: U.S. Pat. 3, 676, 555 (1972).

Manuscript received September 4, 1973; accepted October 17, 1973.

New aspects of organophosphorus pesticides. III. 3-(Thiophosphorylimino-)-oxazolidines, -thiazolidines, -thiazolines, and related compounds as insecticides and acaricides.

By

E. J. Gaughan* and J. J. Menn**

Contents

I. Introduction

The synthesis and insecticidal properties of a new series of phosphorus-containing imino oxazolidines and related compounds were first described by Gaughan (1973). Of special interest in this series of compounds was their excellent systemic insecticidal and acaricidal activity via root uptake from treated soils. Although several organophosphorus and carbamate esters have been developed as systemic insecticides and acaricides, only a few possess a broad spectrum of activity and sufficient persistence in plant tissues to remain active for the susceptible stages of the crop plant during the growing season.

In the oxazolidine imine series one candidate compound, 2-[methoxy(methylthio)-phosphinylimino]-3-ethyl-5-methyl-1,3-oxazolidine (R-16661), showed outstanding activity as a broad-spectrum systemic cotton insecticide and acaricide with useful persistence during the critical growth stages of the cotton plant (Reynolds 1969). This paper reports on the synthesis, structure-activity relations, and

* Stauffer Chemical Co., Western Research Center, Richmond, California 94804.

** Stauffer Chemical Co., Agricultural Research Center, Mountain View, California 94040.

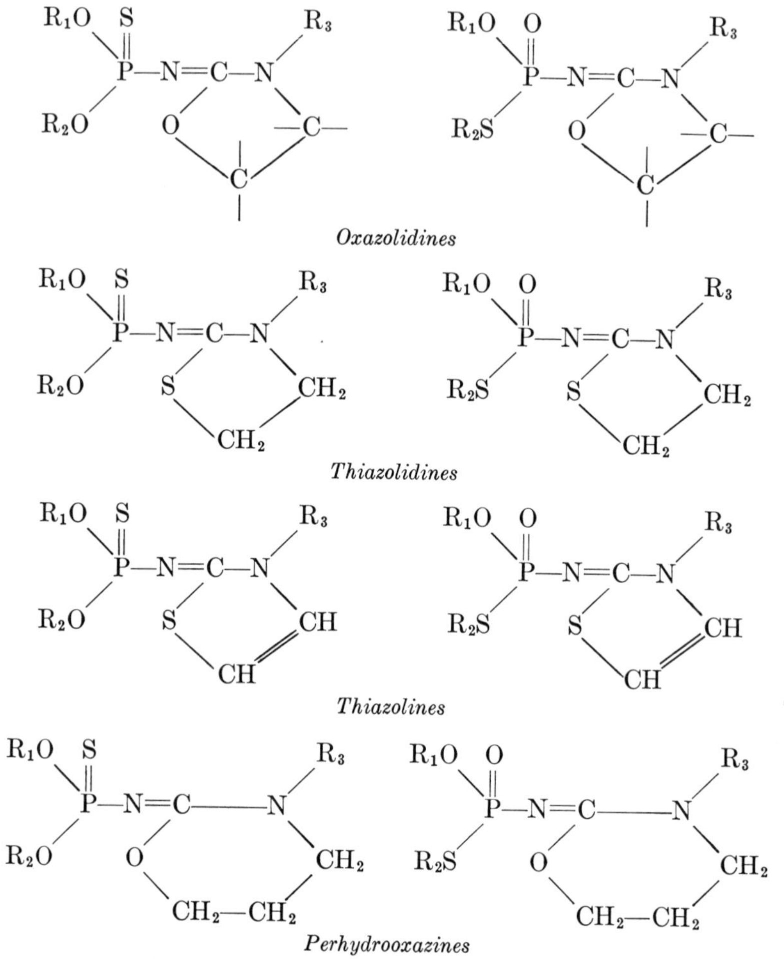

Fig. 1. 3-(Thiophosphorylimino)-oxazolidines, -thiazolidines, -thiazolines, and -perhydrooxazines.

other biological properties of 3-(thiophosphorylimino)-oxazolidines, -thiazolidines, -thiazolines, and -perhydrooxazines (Fig. 1).

II. Synthetic procedures

Two methods of synthesis were employed in preparing the thiophosphorylimino oxazolidines. In the first, a thionochloridate was converted to the isothiocyanate by reaction with potassium thiocyanate in acetonitrile. This was then condensed with the appropriate amino

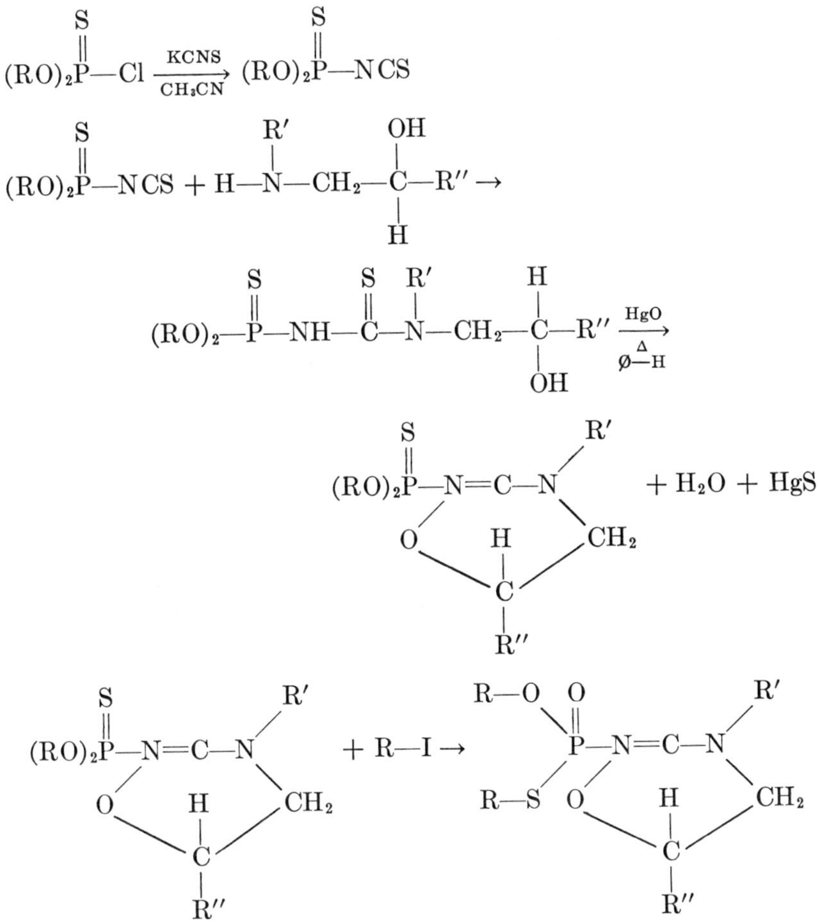

R,R′,R″ = lower alkyl

Fig. 2. Synthesis of thiophosphorylimino oxazolidines.

alcohol. The resulting thiophosphoryl hydroxyalkyl thiourea was then cyclized by reaction with mercuric oxide in refluxing benzene to a thionophosphorylimino oxazolidine. Rearrangement to the thiolo isomer was effected by treatment with alkyl iodide (BURN and CADOGAN 1961), as shown in Figure 2.

The second reaction sequence involved condensation of an amino alcohol with cyanogen bromide in the presence of sodium hydroxide. The resulting hydroxyalkyl cyanamide spontaneously cyclized to a 3-iminooxazolidine (THURSTON 1961). This was then condensed with the thionochloridate and the product rearranged as in the first method, as shown in Figure 3.

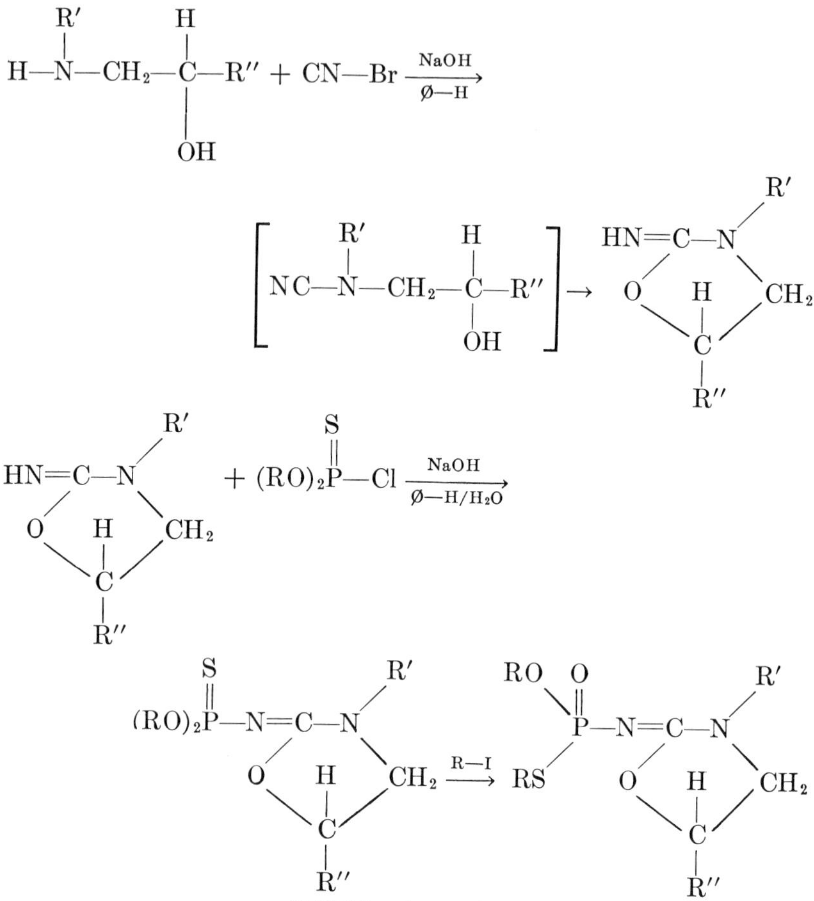

R,R',R'' = lower alkyl

Fig. 3. Second synthetic route in oxazolidine series.

Reaction of the iminooxazolidine with the dichloridate followed by reaction with mercaptan in the presence of sodium hydroxide provided a route to dithiophosphoryl derivatives as indicated in Figure 4.

Optically active 4-alkyl analogs of the oxazolidines were prepared by acetylation of the desired amino acid followed by reduction of the amide acid with LAH to an optically-active amino alcohol. This was used to synthesize optically active 4-alkyl analogs by the usual route as illustrated in Figure 5.

Reaction of an aminothiazoline with an alkyl iodide in methanol produced the ring-alkylated iminothiazolidine iodide salt depicted. Thiophosphorylation in the presence of two equivalents of base fol-

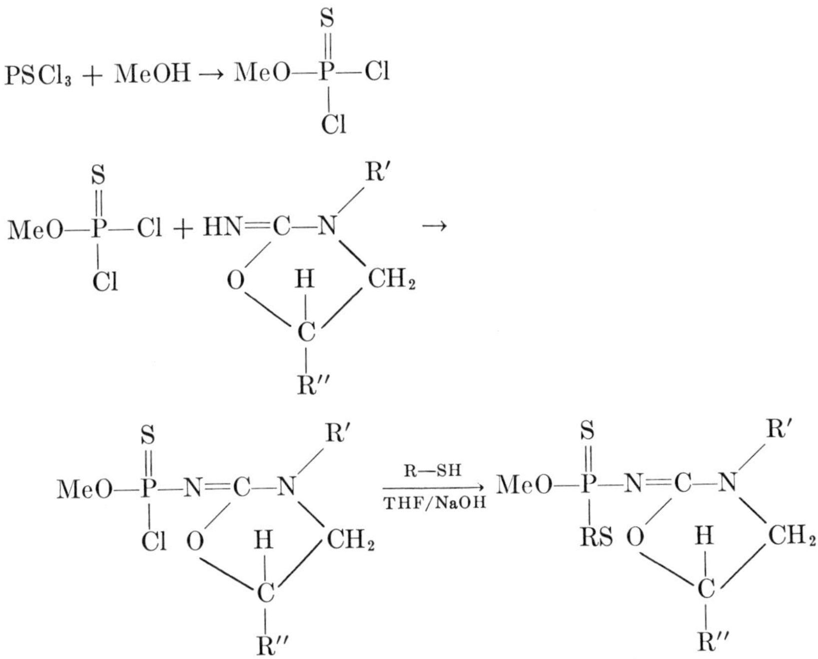

Fig. 4. Synthesis of dithiophosphoryl derivatives.

lowed by rearrangement provided the thiolophosphorylimino thiazolidines. This reaction sequence is depicted in Figure 6.

Condensation of a primary amine with ethyl acrylate and reduction of the amino ester with LAH provided a 3-hydroxypropyl alkyl amine. The cyanamide formed by reaction with cyanogen bromide in the presence of base was cyclized by heat to an iminoperhydrooxazine. The thiolophosphoryl derivatives were prepared as usual. Figure 7 indicates this procedure.

III. Screening procedures

Toxicity tests on houseflies (HF), *Musca domestica* L., were performed on a susceptible strain (S-Stauffer). Twenty-five 3-day-old, caged, female houseflies were exposed for 48 hours to film residues of the toxicant, which was dissolved in an acetone solution containing 0.2 percent peanut oil, on glass Petri dishes (18.8 cm^2). LD_{50} values were determined from log concentration-probit curves. Fourth nymphal instar of the large milkweed bug (MWB), *Oncopeltus fasicatus* (Dallas), and lygus bug adults (LB), *Lygus hesperus* (Knight), were sprayed with aqueous suspensions of the candidate compounds, mortality counts were made 72 hours after treatment,

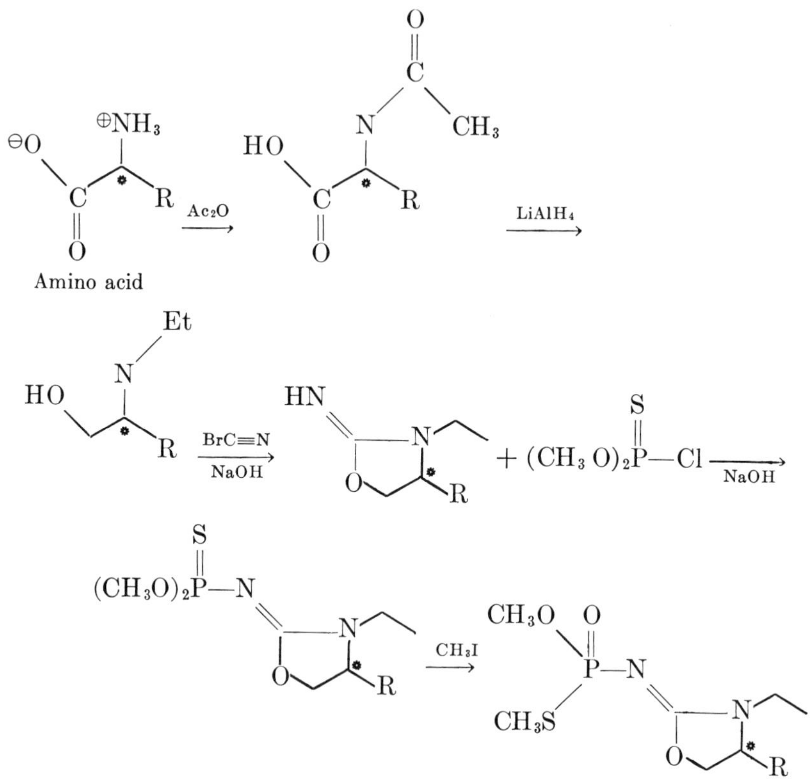

Fig. 5. Synthesis of optically active 4-alkyl analogs.

and LD_{50} values were determined by standard procedures. Third instar larvae of the salt-marsh caterpillar (SMC), *Estigmene acrea* (Drury), and the cotton bollworm (BW), *Heliothis zea* (Boddie), were exposed to sections of treated foliage in Petri dishes containing moistened filter paper discs. Mortality of larvae was determined after 72 hours and LD_{50} values were determined using standard procedures.

Systemic tests with the salt-marsh caterpillar (SMC Sys.) were conducted using kidney bean plants (*Phaseolus vulgaris*) which were immersed in 225-ml wide-mouth glass bottles containing the aqueous test suspension and supported by a cotton plug. Masses of maturing SMC eggs were fastened to the leaves. After seven days, mortality of newly hatched larvae was recorded and LD_{50} values were obtained by standard procedures. Contact and systemic tests with the black bean aphid (BA), *Aphis fabae* (Scop.), were performed using aphid-infested nasturtium plants (*Tropaeolum* sp.). Plants were sprayed with aqueous suspensions of the toxicant for contact activity (BA); for systemic activity (BAS) compounds were incor-

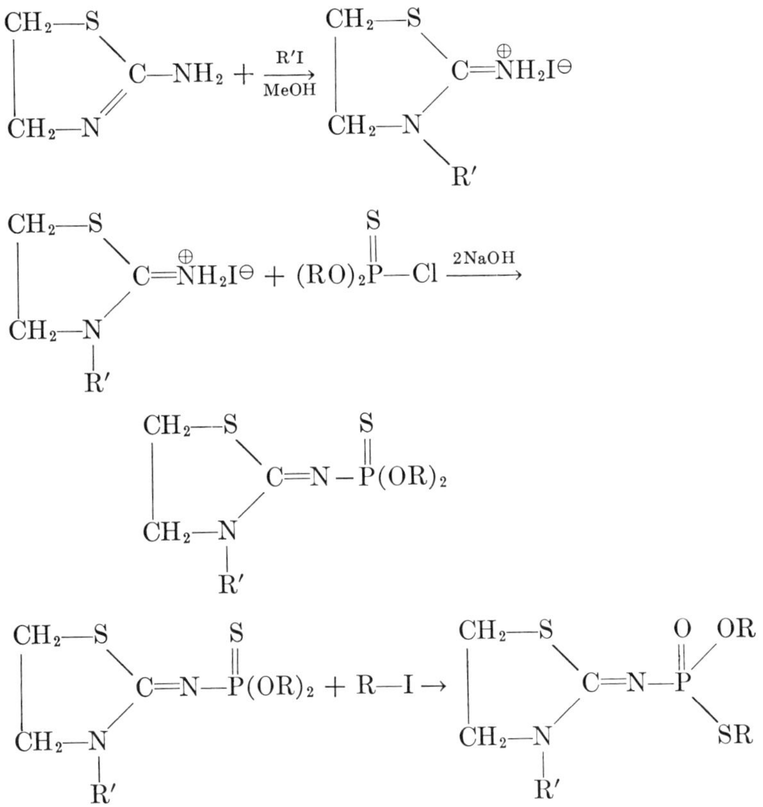

Fig. 6. Preparation of thiophosphorylimino thiazolidines.

porated into the potting soil. Aphid mortality was recorded after 72 hours and LD_{50} values were determined by standard procedures.

Contact and systemic toxicity was also determined using the two-spotted mite (2SM), *Tetranychus urticae* (Koch), using procedures similar to those used with aphids but utilizing the pinto bean (*Phaseolus* sp.) as the host plant and determining mortality after seven days' holding in the greenhouse.

In all spray tests LD_{50} values are expressed as percent toxicant in the spray suspension. Systemic test results are expressed as ppm of toxicant in the aqueous suspension or in the soil. Housefly LD_{50} values are expressed as μg of toxicant/25 female flies.

IV. Structure-activity relationships

Highest level of general systemic insecticidal activity was attained with the thiolo compounds in the oxazolidine series (Table I). The

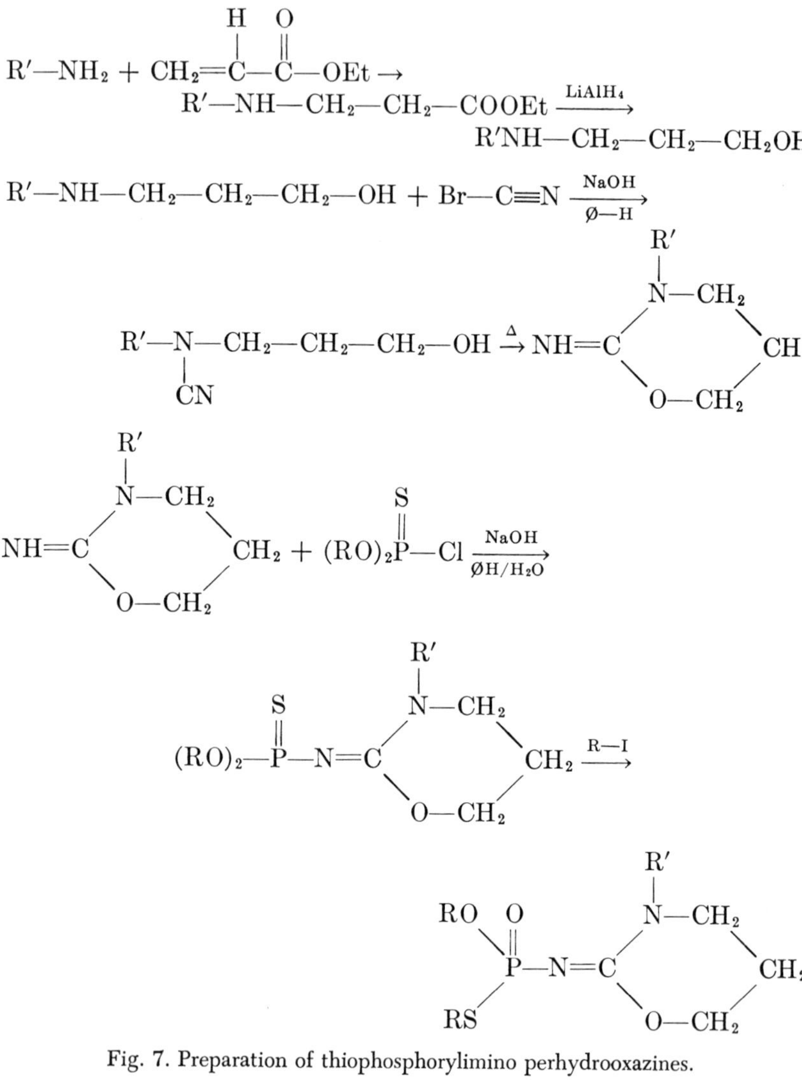

Fig. 7. Preparation of thiophosphorylimino perhydrooxazines.

O,S-dimethyl analog (R-16661) has an acute oral toxicity to rats of 0.12 mg/kg and an acute dermal toxicity to rabbits of 1.5 to 2.0 mg/kg (Kamienski 1969). In general, the monothiolo analogs were more insecticidally active than the corresponding dithio analogs.

A structure-activity study was undertaken in which the size of the P-S-alkyl group (R_1) was progressively increased. The results of this study are in Table II. Increasing the size of the P-S alkyl group (R_1) resulted in lowering of the mammalian toxicity. This, how-

Table I. *Insecticidal activity of thiophosphorylimino oxazolidines.*

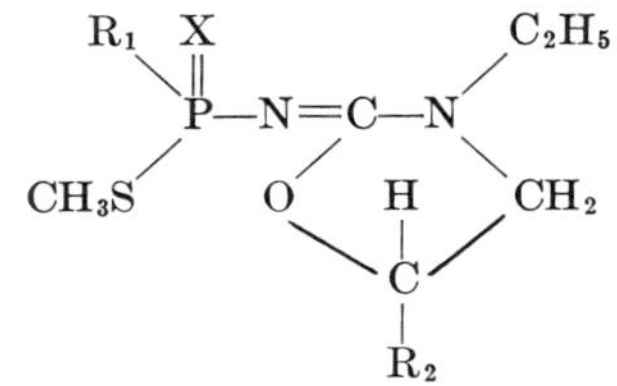

Substituents			LD$_{50}$ for test organism[a]								
R_1	X	R_2	HF (μg/25+)	MWB (%)	LB (%)	SMC (%)	BW (%)	BA (%)	BAS (ppm)	2SM (%)	2SM Sys. (ppm)
CH_3O	O	CH_3	3	0.003	0.0003	0.1	0.005	0.00008	0.008	0.00008	0.01
C_2H_5	O	CH_3	10	0.005	0.0008	0.03	0.03	0.0003	0.03	0.0003	0.3
CH_3O	O	H	5.5	0.005	0.001	>0.1	—	0.0001	0.03	0.0005	—
C_2H_5	S	CH_3	>5	—	0.003	—	—	0.0005	0.8	0.0003	0.3

[a] See text for identification.

Table II. *Effect of phosphorus substituents on biological activity in the oxazolidine series.*

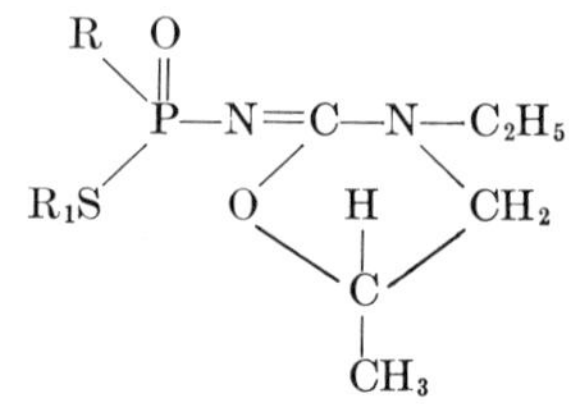

Substituents		LD$_{50}$ for test organism[a]			
R	R_1	BAS (ppm)	2SM Sys. (ppm)	SMC Sys. (ppm)	Rat acute oral (mg/kg)
MeO (R-16661)	Me	0.008	0.03	3	0.1
MeO	Et	0.3	0.3	10	>1
MeO	*n*-Pr	1	3	—	>1
MeO	*i*-Pr	3	8	—	>1
MeO	*n*-Bu	8	inactive	—	—
EtO	Me	0.05	0.4	8	
Et	Me	0.03	0.3	—	<1

[a] See text for identification.

ever, was also accompanied by a much greater decrease in insecticidal activity. The last compound in the table, a thiolophosphonamide, although less insecticidally active than R-16661, proved to have an even higher mammalian toxicity especially in acute dermal tests.

The effect of the ring nitrogen substituent on activity and toxicity was also investigated. Results are summarized in Table III. Again, increasing the size of the alkyl group on the ring nitrogen resulted in lowering of systemic insecticidal activity. In this series, mammalian toxicity remained extremely high until a rather bulky group (*i*-Bu) is introduced.

Several compounds were also prepared with two alkyl substituents on the ring position adjacent to oxygen. Lowered toxicity and activity resulted in these cases.

Another series of oxazolidine derivatives was investigated in which alkyl groups were introduced adjacent to the ring nitrogen atom. In the case of these compounds, an optically active center was maintained. This made possible an evaluation of the effect of optical isomerism on toxicity, cholinesterase inhibition, and insecticidal action. The compounds made and their biological activity are shown in Table IV. Again, it is evident that introduction of a fairly bulky alkyl substituent resulted in a significant loss of systemic insecticidal activ-

Table III. *Effect of variations in ring-alkyl group on activity in the oxazolidine series.*

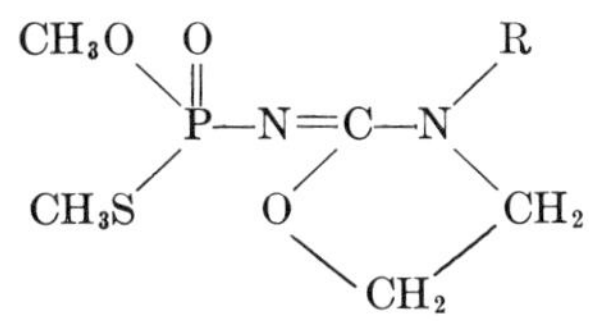

Substituent (R)	LD$_{50}$ for test organism[a]			
	BAS (ppm)	2SM Sys. (ppm)	SMC Sys. (ppm)	Toxicity screen (mg/kg)
Et	0.008	0.03	3.0	<1
i-Pr	0.02	0.04	8	<1
Allyl	0.08	0.03	3	<1
n-Bu	0.5	0.2	3	<1
i-Bu	0	3	—	>1

[a] See text for identification.

Table IV. *Activity of optical isomers in the oxazolidine series.*

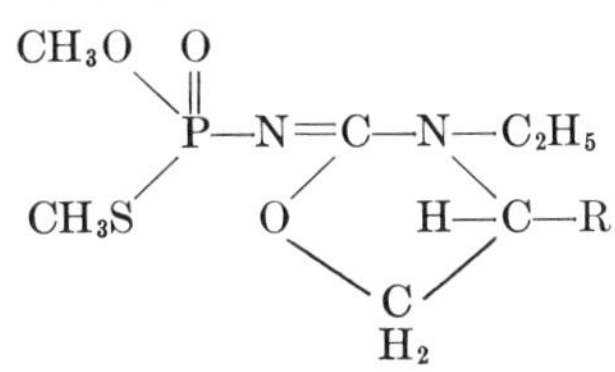

Substituent (R)	LD$_{50}$ for test organism[a]					T$_{50}$[b] (min.)
	HF (μg/25 flies)	BAS (ppm)	2SM Sys. (ppm)	SMC Sys. (ppm)	Toxicity screen (mg/kg)	
CH$_3$ D	6	0.03	0.05	3	1.7	0.8
CH$_3$ L	11	0.08	0.3	3	0.79	0.3
i-Pr L	>100	>10	—	—	>1	4.0
i-Bu L	>100	inactive	inactive	—	—	2.9
R-16661	3	0.008	0.01	0.1	0.12	0.2

[a] See text for identification.
[b] Unpublished data, MIAULLIS and LEE (1969); see text for details of these cholinesterase assays.

ity. However, the methyl compounds retained a good level of potency. With the methyl compound, an approximate ten-fold reduction in mammalian toxicity compared to the racemic 5-methyl analog (R-16661) was realized. However, even in this case the toxicity remained surprisingly high.

The cholinesterase inhibition assays (Table IV) were determined using housefly (HF) head suspensions (one g/100 ml) to which aliquots of the inhibitor solution were added to make the final inhibitor concentration $1 \times 10^{-0}M$. The mixture was incubated at 37°C and at different time intervals a 20-μl aliquot was taken to determine the residual enzyme activity in a cuvette containing three ml of 1.0 m*M* acetylthiocholine and 1.26 M*M* DTNB (5,5′-dithio-bis-2,2′-nitrobenzoic acid; Aldrich). Measurements were made at 415 nm with a Unicam SP800 spectrophotometer. The time required to inhibit 50 percent of the initial enzyme activity (T_{50}) was obtained from plots of residual enzyme activity vs. incubation time. The T_{50} values shown in Table IV indicate that rapid inhibition of HF head enzyme generally correlated with increased toxicity to insects and the rat. Larger substituents (R) contributed to reduction in activity while the optical isomers had no significant effect on changing toxicity and enyzyme inhibition.

Expansion of the five-membered oxazolidine ring to a six-membered perhydroxazine system had the effect of reducing insecticidal activity without reduction in mammalian toxicity.

The thiazolidine derivatives were an interesting class of compounds with a somewhat different spectrum of activity These results are summarized in Table V.

The first compound in Table V, the *O,S*-dimethyl *N*-methyl thiolophosphate, was the most active. It displayed high contact and systemic toxicity to the bean aphid, but the compound was not an effective acaricide. However, the salt marsh caterpillar systemic activity was as good as that of the best of the thiolo oxazolidine derivatives. The acute oral toxicity of 1.5 mg/kg (LD_{50}) to female rats was about ten-fold less than that of the most insecticidally active oxazolidine. An acute rabbit dermal LD_{50} of 3.16 mg/kg was found for this material. Consequently, the compound must still be considered highly toxic to mammals.

The unsaturated thiazolidines exhibited significantly lower activity than the thiazolidines as indicated in Table VI. However, introduction of an *N*-propargyl group did increase systemic spider mite activity.

V. Metabolism

Fukuto and others (FUKUTO *et al.* 1972) studied the metabolism of R-16661 in cotton plants, houseflies, and in *in vitro* enzymatic model oxidation systems. They found that it was quite stable in cotton

Table V. *Activity of thiophosphoryliminothiazolidines.*

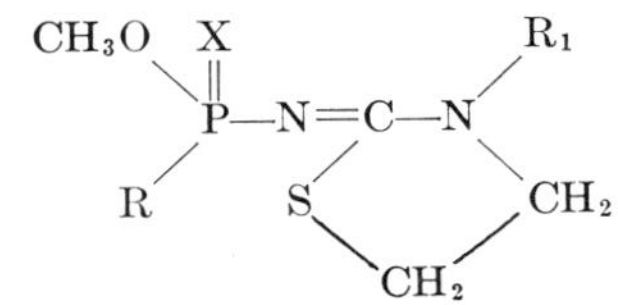

Substituents			LD$_{50}$ for test organism[a]						
R	X	R$_2$	Toxicity screen (mg/kg)	HF (μg/25 flies)	BA (%)	BAS (ppm)	2SM PE (%)	2SM Sys. (ppm)	SMC Sys. (ppm)
CH_3S	O	CH_3	1.5	15	0.003	0.03	0.005	3	3
CH_3S	O	C_2H_5	>1	20	0.008	0.3	0.03	5	8
CH_3O	S	C_2H_5	—	100	0.03	3	>0.05	>10	—
CH_3O	S	n-C_3H_7	—	>100	>0.05	—	>0.05	>10	>10
CH_3S	O	CH_2—$CH{=}CH_2$	—	50	0.003	0.3	0.05	3	>10
CH_3O	S	CH_2—$CH{=}CH_2$	—	>100	0.05	3	>0.05	>10	—
CH_3O	S	CH_2—$CH{=}CCl_2$	—	50	0.03	3	>0.05	3	>10
CH_3O	S	n-C_5H_{11}	—	>100	>0.05	—	>0.05	>10	>10

[a] See text for identification.

Table VI. *Activity of thiophosphoryliminothiazolines.*

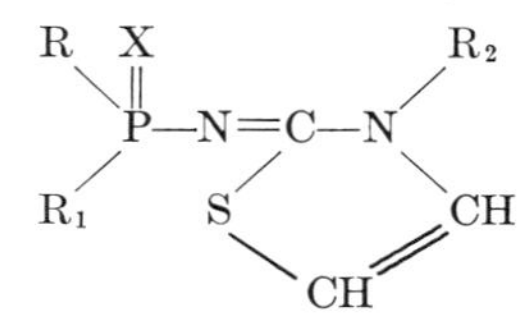

Substituents				LD$_{50}$ for test organism[a]						
R	R$_1$	X	R$_2$	Toxicity screen (mg/kg)	HF (μg/24 flies)	BA (%)	BAS (ppm)	2SM PE (%)	2SM Sys. (ppm)	SMC Sys. (ppm)
CH_3O	CH_3O	O	CH_3	>1	50	0.003	0.8	0.03	3	>10
C_2H_5O	CH_3S	O	CH_3	—	45	0.03	3	0.01	8	>10
CH_3O	CH_3S	O	CH_2—CH≡CH	—	10	0.03	3	0.03	0.8	>10
CH_3O	CH_3S	O	C_2H_5	—	—	0.01	3	0.03	3	—
CH_3O	CH_3O	S	C_2H_5	—	>100	0.05	5	0.05	3	>10

[a] See text for identification.

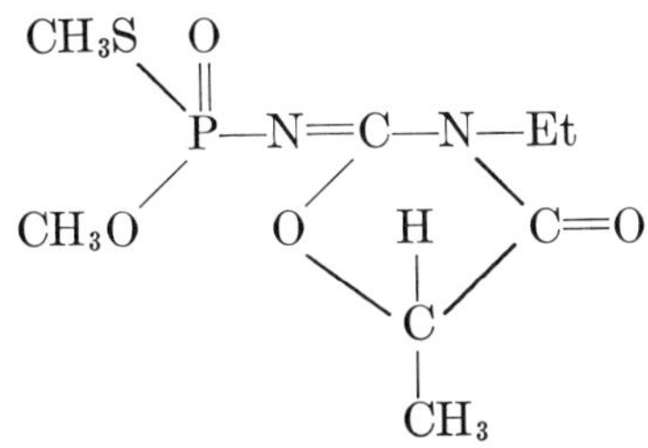

Fig. 8. 4-Keto metabolite of R-16661.

plants and houseflies. It was slowly metabolized to give the 4-keto derivative shown in Figure 8. This metabolite proved to be much less insecticidal to houseflies than R-16661 itself, but it was of almost equal toxicity to the white mouse. These results suggest that R-16661 itself and not some metabolite thereof is the actual toxicant in insects.

Summary

A series of 3-(thiophosphorylimino)-oxazolidines, -thiazolidines, -thiazolines, and -perhydrooxazines has been prepared and evaluated for insecticidal and acaricidal activity. Many of these compounds exhibited a high level of biological activity, particularly as systemic insecticides and acaricides. In general, the thiolophosphoryl isomers were more active than the thiono compounds. Studies exploring structure-activity relationships in these series are presented and some results of an investigation on the metabolism of Stauffer R-16661 are given. The highly active insecticides discovered in this work have also proven to be highly toxic to mammalian species.

References

Burns, A. J., and J. I. G. Cadogan: The reactivity of organophosphorus compounds. Part IX. The reaction of thionates with alkyl iodide. J. Chem. Soc. **1961**, 5532.

Fukuto, T. R., S. P. Shrivastava, and A. L. Black: Metabolism of 2-[methoxy-(methylthio)phosphinylimino]-3-ethyl-5-methyl-1,3-oxazolidine in the cotton plant and houseflies. Pest. Biochem. Physiol. **2**, 162 (1972).

Gaughan, E. J.: U.S. Pat. 3,714,176 (1973).

Kamienski, F. X.: Unpublished results (1969).

Miaullis, J. B., and H. Lee: Unpublished results (1969).

Reynolds, H. T.: Personal communication (1969).

Thurston, J. T.: Oxazolidines and a method of preparing the same. U.S. Pat. 2,479,525 (1949).

Manuscript received September 4, 1973; accepted October 17, 1973.

New aspects of organophosphorus pesticides. IV. Newer aspects of the metabolism of phosphonate insecticides.

By

JULIUS J. MENN* and J. BRUCE MCBAIN*

Contents

I. Introduction

Organophosphorus (OP) compounds comprise an increasingly important class of insecticides owing to the greater emphasis on use of more selective and biodegradable chemicals for management of insect pest populations. There are available at present approximately 85 commercial and experimental OP insecticide chemicals (KENAGA and ALLISON 1969) including several phosphonate esters (OP compounds with a P—C bond).

Generalized structural formulae of phosphonate ester insecticides are shown in Figure 1, where R_1 and R_2 are usually similar or dissimilar, aryl, straight, or branched alkyl chains ranging from one-to-five carbon atoms and X, the leaving group, can consist of chloroalkyl, cyclic, or aromatic moieties. The structures of several commercial phosphonate insecticides are shown in Figure 2.

* Stauffer Chemical Co., Agricultural Research Center, Mountain View, California 94042.

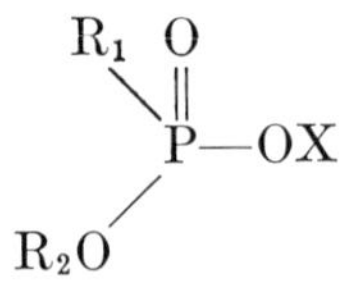

Phosphonate

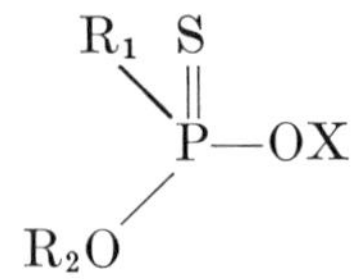

Phosphonothioate

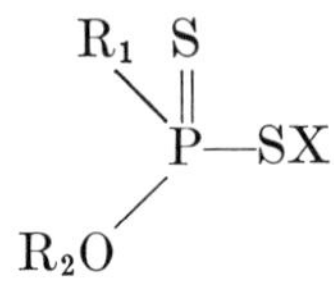

Phosphonodithioate

R_1, R_2 = lower alkyl, chloroalkyl, or phenyl
X = leaving group

Fig. 1. Generalized structural formulas of organophosphonate insecticides.

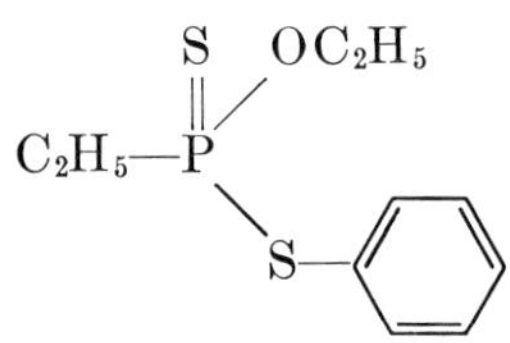

Dyfonate

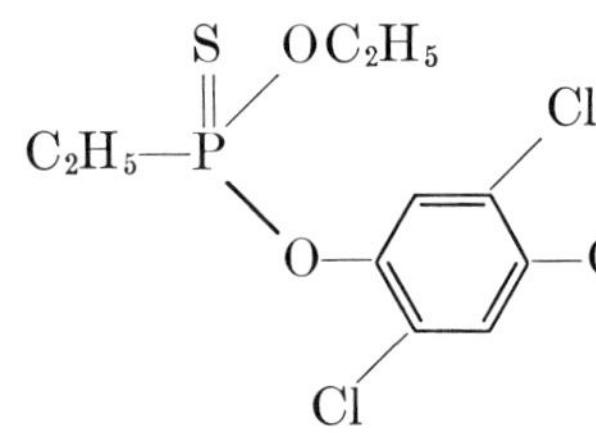

Agritox

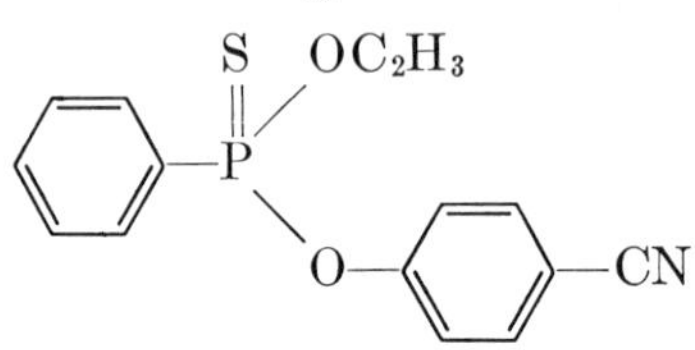

Surecide

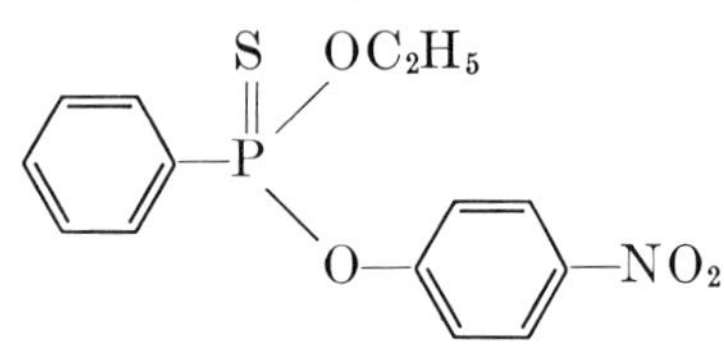

EPN

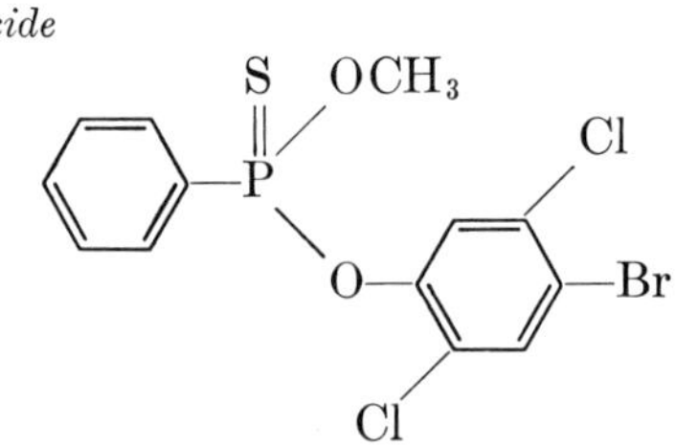

Leptophos

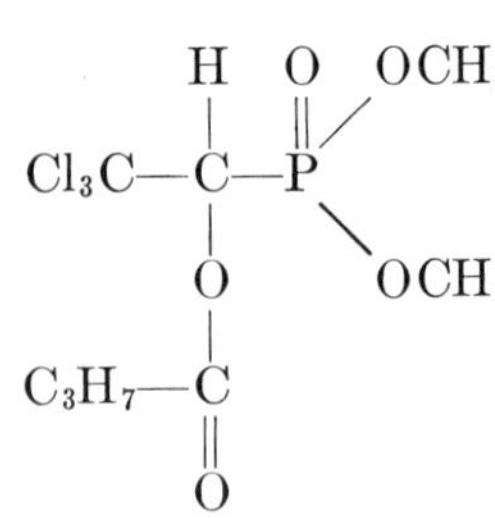

Butonate

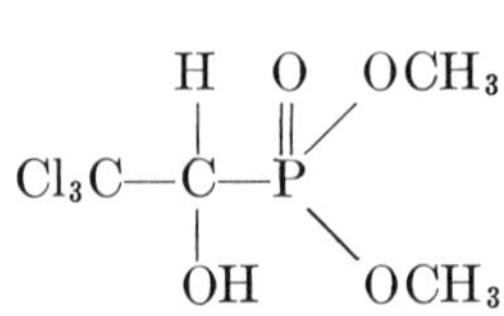

Trichlorfon

Fig. 2. Structures of commercial phosphonate insecticides.

The first phosphonate ester, the diethyl ester of methylphosphonic acid, was synthesized in 1873 by A. V. Hofmann who recognized also the high chemical stability of the P—C bond (SCHRADER 1967). These early studies provided the chemical background which led to the synthesis in 1938 of the potent cholinesterase inhibitor, Sarin (*O*-isopropyl fluoro methylphosphonic acid), a forerunner of the phosphonate insecticides (SCHRADER 1963).

The metabolic fates of phosphonate insecticides in plants and animals, with special attention to terminal residues, were the subject of a recent review (MENN 1971). The current review updates what is known concerning the metabolism of phosphonates with emphasis on the fate of the phosphorus moiety using as examples primarily insecticides and certain other compounds. Additionally, attention is focused on the differences and similarities between phosphonates and phosphates using comparative metabolism and biological activity as a basis for discussion.

II. Metabolism

a) Chloroethylphosphonates

The insecticides trichlorfon (*O,O*-dimethyl 2,2,2-trichloro-1-hydroxyethylphosphonate) and Butonate (*O,O*-dimethyl 2,2,2-trichloro-1-butyryloxyethylphosphonate) and the plant-growth regulator ethephon (2-chloroethylphosphonic acid) are in a sense anomalous phosphonates because the P—C bond involves the "leaving group." Trichlorfon and Butonate readily undergo rearrangement *in vitro* at pH 6 or higher to their respective phosphate forms, and in animals and plants the metabolites produced *in vivo* are, in part, derivatives of phosphoric acid (MENN 1971). Ethephon-^{14}C, in alkaline solution (YANG 1969) or applied to plants (YAMAGUCHI *et al.* 1971), cleaved to yield ethylene. The probable chemical mechanism of ethylene production as reviewed by YANG (1969) involves a nucleophilic attack at the phosphorus by a water or hydroxide ion and the concerted elimination of chlorine leading to formation of ethylene, chloride, and phosphate.

b) Methylphosphonates

The metabolites formed from methylphosphonates in animals and plants and pathways leading to their formation are presented in Table I and Figure 3.

The fate of Soman (*O*-pinacolyl methylphosphonofluoridate) was studied in wheat plants by HAMBROOK *et al.* (1971) and *in vitro* using rat tissue preparations by HARRIS *et al.* (1964). Young wheat plants grown for 24 hours in a hydroponic culture medium containing 20 to 25 ppm of ^{32}P-Soman were extracted and the radiolabeled prod-

Table I. *Metabolites of phosphonates formed by mammals and plants.*[a]

Compound[b]	Metabolites isolated from urine (% urinary radioactivity) or from plants (% total radioactivity)[b,c]													
	$R_1P(S)(OR_2)(OH,SH)$		$R_1P(O)(OR_2)(OH,SH)$		$R_1P(O)(OH)(OH,SH)$		$R_1P(O)(OR_2)(X)$		$R_1P(S)(OH)(X)$		$R_1P(O)(OH)(X)$		Others[d]	
	P	A	P	A	P	A	P	A	P	A	P	A	P	A
Methylphosphonates														
I Methylparaphonothion (mouse)	—	7.5	—	60.5	—	21.3	—	6.4	—	0.9	—	1.2	—	U-2.2
II Sumiphonothion (mouse)	—	4.2	—	57.5	—	24.9	—	7.7	—	1.2	—	2.6	—	U-1.9
III Soman (wheat)	—	—	95.7	—	4.3	—	0	—	—	—	?	—	0	—
Ethylphosphonates														
IV Dyfonate (rat, potato)	2.9	60.1	51.6	36.1	—	0	3.4	0.2	—	0	—	0	B-26.5 P- 0.5 U-15.1	P-0.2 U-3.4
V N-4543 (rat)	—	3.0	—	92.0	—	0	—	0	—	0	—	0	—	U-5.0
Phenylphosphonates														
VI Leptophos (mouse, cotton)	7.0	62.7	2.1	19.8	10.3	17.2	trace	0	0	0	0	0.2	B-11.8 P-68.8	0
VII Inezin[e] (rice)	—	—	21.0	—	30.3	—	—	—	—	—	11.6	—	P- 4.4 U-17.7 EPTA-15.0	—

[a] Data from ENDO *et al.* (1970) (VII), HAMBROOK *et al.* (1971) (III), HOLLINGWORTH *et al.* (1967) (I and II), HOLMSTEAD *et al.* (1973) (VI), MCBAIN *et al.* (1970 and 1971) (IV), and MENN (1971) (V).

[b] R_1 = —CH_3 (I, II, and III), —C_2H_5 (IV and V), —Ø (VI and VII); R_2 = —CH_3 (I, II, and VI), —$CH(CH_3)C(CH_3)_3$ (III), —C_2H_5 (IV),—C_4H_9-*i* (V), and —C_2H_5 (VII); X=

—OØ—4NO_2 (I), —OØ—3CH_3—4NO_2 (II), —F (III), —SØ (IV), —SCH_2N 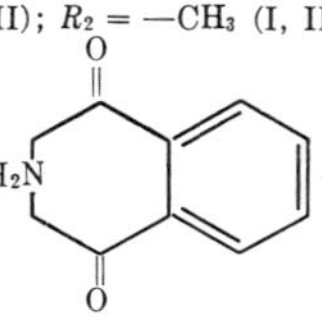 (V), —OØ—4Br—2, 5Cl_2 (VI), and —SCH_2Ø (VII).

[c] P = plant and A = animal.

[d] B = bound, P = parent compound, U = unknown, and $EPTA$ = $(C_2H_5O)P(O)(OH)(SH)$.

[e] Metabolite distribution expressed as percent of radioactivity in a methanolic extract which contained about 15 percent of the total radioactivity.

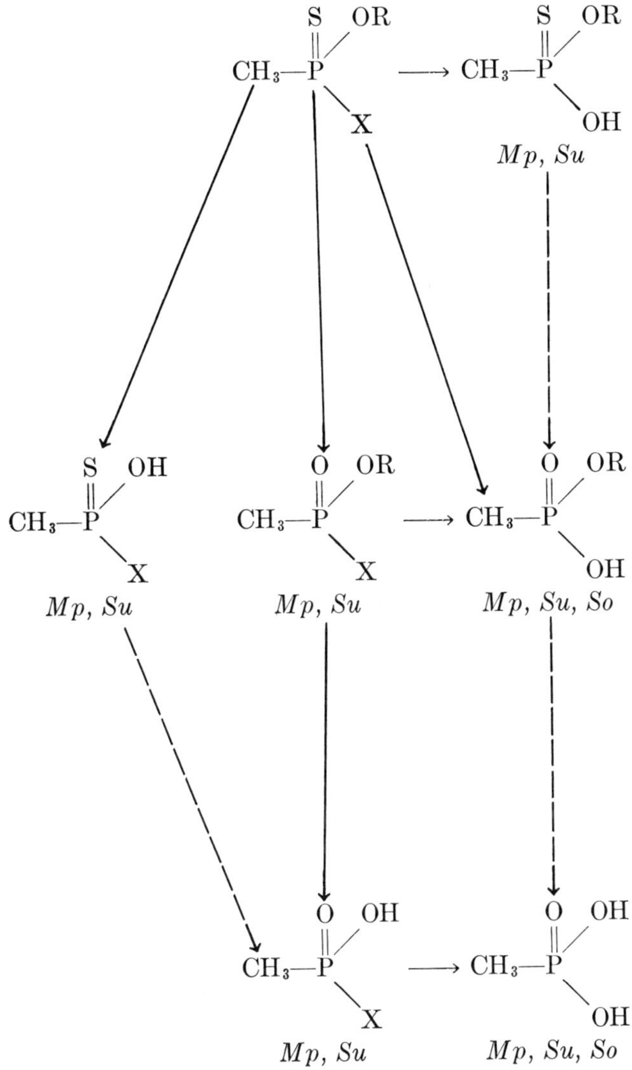

	R	X
Mp	$—CH_3$	$—OØ—4NO_2$
Su	$—CH_3$	$—OØ—3CH_3—4NO_2$
So	$—CH(CH_3)C(CH_3)_3$	$—F$

Fig. 3. Generalized metabolic pathways for methylphosphonates in animals (*A*) and plants (*P*) *in vivo:* Methyl paraphonothion (*Mp*) and (*A*) (HOLLINGWORTH *et al.* 1967), Sumiphonothion (*Su*) and (*A*) (HOLLINGWORTH *et al.* 1967), and Soman (*So*) and (*P*) (HAMBROOK *et al.* 1971).

ucts thus recovered were analyzed by thin-layer chromatography (TLC) and gas-liquid chromatography (glc) after methylation. Soman hydrolyzed by cleavage of the P—F bond both in the culture medium and within the plant to pinacolyl methylphosphonic acid and the latter was further degraded by the plant to methylphosphonic acid. No evidence for P—C bond cleavage was obtained. Soman-^{32}P, when incubated with the plasma or the 40,000 x g supernatant of a rat-liver homogenate, was shown by paper chromatographic analyses to decompose readily to pinacolyl methylphosphonic acid. Further breakdown to methylphosphonic acid did not occur.

The metabolic fate of the experimental insecticide Colep-phenyl-^{14}C [*O*-phenyl *O*-(4-nitrophenyl) methylphosphonothionate] was studied in the rat and plants by MARCO and JAWORSKI (1964). Limited desulfuration of Colep to the oxon was indicated in plants. The major radioactive plant and rat urinary metabolites appeared to be conjugates of the cleaved phenol. These findings imply that *O*-(4-nitrophenyl) methylphosphonic and phosphonothioic acids were also formed as metabolites. Additionally, cleavage of the nitrophenyl moiety is also likely since it is a better leaving group than the phenol. Cleavage of both phenolic groups would give rise to methylphosphonic acid as the terminal metabolite, keeping the P—C bond intact.

HOLLINGWORTH *et al.* (1967) studied the metabolism in mice of methyl parathion and Sumithion and their phosphonate analogs which they named methyl paraphonothion and Sumiphonothion, respectively. Mice were orally dosed with 2.5 to 850 mg/kg of the ^{32}P-labeled insecticides and the urine and feces were collected separately up to 72 hours after dosing. Urinary metabolites were separated and identified by ion-exchange and paper chromatography in comparison with reference standards.

Excretion of urinary ^{32}P was slower in rats treated with the phosphonothioates than from rats treated with the corresponding phosphorothioates. Metabolites arising from the administered insecticides were identified as the dimethyl thiophosphorus, dimethyl phosphorus, and methyl phosphorus acids, the oxons, the demethyl analogs of the parent compounds and their oxons, and from the phosphates only, phosphoric acid. Small amounts of ^{32}P from each compound remained unidentified. There was no indication for cleavage of the P—C bonds of the phosphonates.

Table II compares the urinary metabolites excreted during the first 24 hours after dosage with 2.4 to 3.1 mg/kg of methyl parathion and Sumithion and their respective phosphonate analogs. Metabolism of the phosphonates differs in several respects from the corresponding phosphorothionates. One outstanding aspect of phosphonate metabolism is the relatively high proportion of total P=O metabolites which, the authors suggested, indicates that phosphonates are subject to more extensive oxidation than the analogous phosphates. There are data

Table II. *Comparative metabolism of analogous phosphonate-phosphate insecticide pairs in the mouse.*[a]

Compound	Metabolites isolated from urine (% total urinary activity)[b]							
	R–P(=S)(OCH$_3$)(OH)	R–P(=O)(OCH$_3$)(OH)	R–P(=O)(OH)(OH)	HO–P(=O)(OH)(OH)	R–P(=O)(OCH$_3$)(X)	R–P(=S)(OH)(X)	R–P(=O)(OH)(X)	Unknowns
I. Methyl parathion	12.9	31.9	2.0	5.8	2.4	18.8	23.1	3.1
II. Methyl paraphonothion	7.5	60.5	21.3	0	6.4	0.9	1.2	2.2
III. Sumithion	20.3	21.4	2.5	2.4	1.6	20.1	28.4	2.5
IV. Sumiphonothion	4.2	57.5	24.9	0	7.7	1.2	2.6	1.9

[a] From HOLLINGWORTH *et al.* (1967).
[b] R = —CH_3 (II and IV) or —OCH_3 (I and III), X = —OØ—4NO_2 (I and II) or —OØ—3CH_3—4NO_2 (III and IV).

to indicate that certain phosphonothionates, upon exposure to air while deposited upon the surface of cellulose chromatoplates, are more readily converted, probably by oxidation, to cholinesterase inhibitors than their respective phosphorothionate analogs (MCBAIN and MENN 1964). HOLLINGWORTH *et al.* (1967) further suggested that phosphonothionates are less readily cleaved and thus accumulate at the site of oxidation which helps to explain the higher toxicity of these compounds and the slower elimination of their urinary metabolites. Another difference noted in these studies indicates that the intact phosphonates undergo considerably less extensive desalkylation than the phosphates or, perhaps as indicated by the much greater accumulation of monoalkylphosphorus acid, the phosphonate desalkylation products are more readily degraded and therefore are less likely to accumulate.

c) *Ethylphosphonates*

The products arising from the metabolism of ethylphosphonates in plants and animals and the pathways leading to their formation are shown in Table I and Figure 4.

The soil insecticide Dyfonate (*O*-ethyl *S*-phenyl ethylphosphonodithioate), radiolabeled in the α-carbon of the ethoxy group or the benzene ring, was studied in rats (MCBAIN *et al.* 1971 a) and potato plants (MCBAIN *et al.* 1970). Elimination of Dyfonate-ethoxy-^{14}C from the rat was virtually complete 48 hours following a single oral dose of 2 to 8 mg/kg and occurred principally in the urine (90.7 percent) and feces (7.4 percent) with a trace found in the expired air. Identity of urinary metabolites was established by TLC, glc (acidic products were first methylated), and mass-spectral analyses. They were shown to comprise *O*-ethyl ethylphosphonothioic

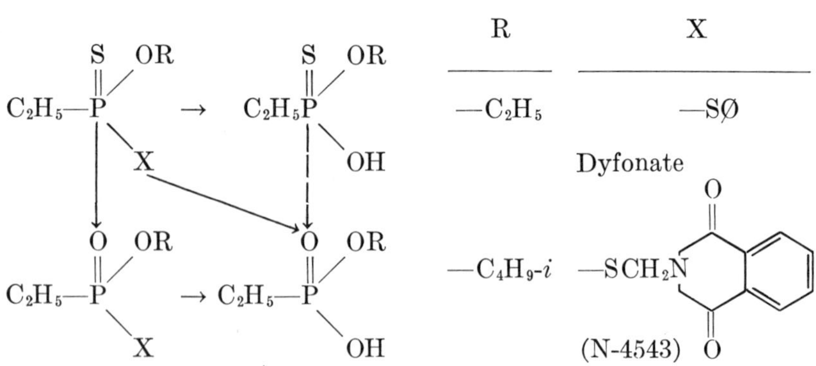

Fig. 4. Generalized metabolic pathway for ethylphosphonates in animals (*A*) and plants (*P*) *in vivo:* Dyfonate (*A* and *P*) (MCBAIN *et al.* 1970 and 1971) and N-4543 (*A*) (MENN 1971).

and *O*-ethyl ethylphosphonic acids and trace amounts of unchanged Dyfonate and its oxon. *O*-Desethylation was not a significant pathway as indicated by the essential lack of $^{14}CO_2$ in the expired air of the rat.

The tissues of potato plants grown for up to 87 days in soil containing 2.64 ppm of Dyfonate-ethoxy-^{14}C yielded the same phosphorus-containing metabolites as found in rat urine. Additionally, a significant fraction of the total radiocarbon was associated in an undefined manner with the insoluble plant residues.

The phosphonic acid metabolites of Dyfonate, with the exception of the oxon, are only slightly toxic to rats orally and to the housefly topically.

Metabolism of the experimental insecticide N-4543 (*S*-phthalimidomethyl *O*-isobutyl ethylphosphonodithioate) radiolabeled with isobutoxy-1-^{14}C or carbonyl-^{14}C was studied in the rat (MENN 1971). Within 96 hours after dosing at 4.5 mg/kg with N-4543-isobutoxy-1-^{14}C, the recovery of administered radioactivity was as follows: 62 percent in urine, 34 percent in feces, and 0.6 percent in expired air. Urinary metabolites, characterized by TLC with reference standards, were shown to be isobutoxy ethylphosphonic and isobutoxy ethylphosphonothioic acids, both retaining the P—C bond, and two unknowns, one of which may possibly represent an acid-labile conjugate (other than the glucuronide or sulfate) of isobutoxy ethylphosphonic acid. It is apparent that the P—O—isobutyl bond remained intact since $^{14}CO_2$ was not a significant metabolite.

d) Phenylphosphonates

Although the arylphosphonothioate insecticide and acaricide EPN (*O*-ethyl *O*-4-nitrophenyl phenylphosphonothioate) was first introduced 25 years ago, almost no published information is available on its metabolism in animals, and none exists for plants. The few published reports deal with biotransformations of EPN in isolated biological systems. NAKATSUGAWA *et al.* (1968) examined the degradation of EPN by the NADPH-dependent microsomal oxidase system of rabbit liver. Oxidative metabolism resulted in the formation of EPN-oxon and cleavage of the acid anhydride bond of EPN, but not of the oxon, as indicated by the release of 4-nitrophenol.

AHMED *et al.* (1958) demonstrated that upon incubation of EPN and its oxon with cow rumen juice, five percent of EPN and 50 percent of the oxon were reduced to their respective nontoxic amino analogs. The remaining material from each starting product was accounted for as cleavage products. The higher recovery of the amino phosphonate was due to the greater rate of cleavage of the phosphonothioate than the phosphonate by rumen juice.

The fate of leptophos [*O*-(4-bromo-2,5-dichlorophenyl) *O*-methyl phenylphosphonothioate] radiolabeled with ^{14}C in the phenyl or phenoxy positions was determined following oral administration to mice at 25 mg/kg and topical application to foliage of greenhouse-grown cotton plants (HOLMSTEAD *et al.* 1973). Radiolabeled products recovered from the urine and feces of mice and from external washes and homogenates of cotton leaves were characterized by TLC and mass spectra. The rate of elimination of radiocarbon from mice was slower for the phenyl label (87.9 percent after 144 hours) than for the phenoxy label (virtually complete after 48 hours). Radiocarbon from the phenyl label was excreted 90 percent in the urine and ten percent in the feces.

The metabolites of leptophos and the proposed pathway leading to their formation are presented in Table I and Figure 5. Unchanged

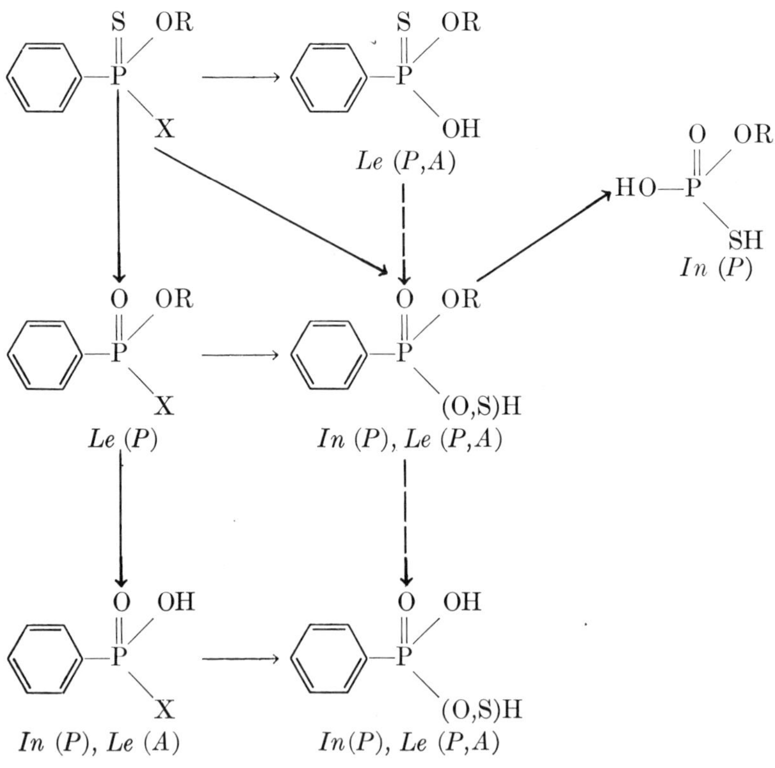

	R	X
Le	—C_2H_5	—$SCH_2Ø$
In	—CH_3	—OØ—4Br—2,5Cl_2

Fig. 5. Generalized metabolic pathway for phenylphosphonates in animals (*A*) and plants (*P*) *in vivo:* Inezin (*In*) (ENDO *et al.* 1970) and leptophos (*Le*) (HOLMSTEAD *et al.* 1973).

leptophos and its oxon were found in small quantities primarily in the feces. The principal metabolites in the urine collected through 144 hours were *O*-methyl phenylphosphonothioic, *O*-methyl phenylphosphonic and phenylphosphonic acids, and trace amounts of a product tentatively identified as the demethyl oxon [*O*-(4-bromo-2,5-dichlorophenyl) phenylphosphonic acid].

Leptophos does not readily penetrate into the leaves of cotton plants but remains on the surface and is primarily lost by volatilization. Qualitatively, the degradation of leptophos in or on cotton leaves is similar to that occurring in the mouse. Little if any oxon was present at any sampling period, the phenylphosphonic acid derivatives were present in and on leaves in varying amounts during the five weeks of sampling. A small but significant amount of radiocarbon of unknown nature was inseparable from the plant pulp and was considered to be bound.

With the exception of the 2-chloroethylphosphonates which readily cleave at the P—C bond, the evidence discussed thus far indicates that the P—C bond of methyl-, ethyl-, and phenylphosphonates resists catabolism by higher animals or plants. A contrary indication, however, is provided by a metabolism study of the fungicide Inezin (*O*-ethyl *S*-benzyl phenylphosphonate) in rice plants (ENDO *et al.* 1970). Radiolabeled products extracted from the plants five days following treatment with Inezin-^{35}S were tentatively identified by TLC co-chromatography with reference standards as unchanged Inezin, desethyl Inezin, *O*-ethyl phenylphosphonothioic acid, phenylphosphonothioic acid, and, most significantly, *O*-ethyl phosphorothioic acid (Table I, Fig. 5). The presence of the latter acid provides the first indication that the P—C bond of a phosphonate other than a chloroethylphosphonate may be catabolized in a higher organism.

III. Mechanisms of metabolite formation

The studies reviewed here have described a number of biotransformation products resulting from the metabolism of phosphonates by plants and animals, but little has been said thus far regarding the mechanisms involved in the formation of these products. The number of published studies describing the mechanisms responsible for metabolism of phosphonate insecticides is quite meager in comparison to those available for organophosphates. However, the same mechanisms are expected to operate in the biotransformation of phosphonate or phosphate insecticides.

Thionophosphorus insecticide chemicals generally are activated to cholinesterase inhibitors by oxidation to their corresponding oxygen analogs (oxons) and are detoxified, most commonly, by cleavage of their acid anhydride bond yielding disubstituted thionophosphorus acids *in vivo* in mammals and other organisms or when incubated with the NADPH-dependent microsomal oxidase system prepared from certain tissues of these organisms (DAUTERMAN 1971, MENN 1971, BULL 1972). These enzymatically mediated reactions require both NADPH and atmospheric oxygen and, therefore, oxidative mechanisms are considered to be involved. It was shown in ^{18}O tracer studies with the phosphonodithioate Dyfonate (MCBAIN *et al.* 1971 b) and the phosphorothionate parathion (PTASHNE *et al.* 1971) (Fig. 6) that in the mammalian hepatic NADPH-dependent microsomal oxidase system the oxons receive their oxygen from molecular oxygen, while the diethyl thiophosphorus acids receive their oxygen from water. Diethyl phosphorus acid, also a metabolite of Dyfonate or parathion formed by the microsome system, receives one oxygen from molecular oxygen and one oxygen from water.

The pattern of ^{18}O incorporation into these metabolites suggests that the microsomal oxidase catalyzed desulfuration and anhydride bond cleavage reactions of phosphoro- and phosphonothioate insecticides proceed by the initial formation of oxygenated intermediates involving the addition of an oxygen atom from molecular oxygen to the insecticide at the thionosulfur. The complex then cleaves to the oxon and an unidentified sulfur metabolite or, alternatively, reacts with water and cleaves with the loss of the introduced molecular oxygen and the leaving group to diethyl thiophosphorus acid. The oxon hydrolyzes by esterase action to yield dialkyl phosphorus acid, although it is possible that this acid may also be formed, in part, directly from the proposed oxygenated intermediate. The diethyl thiophosphorus acids are not further metabolized by the microsome sys-

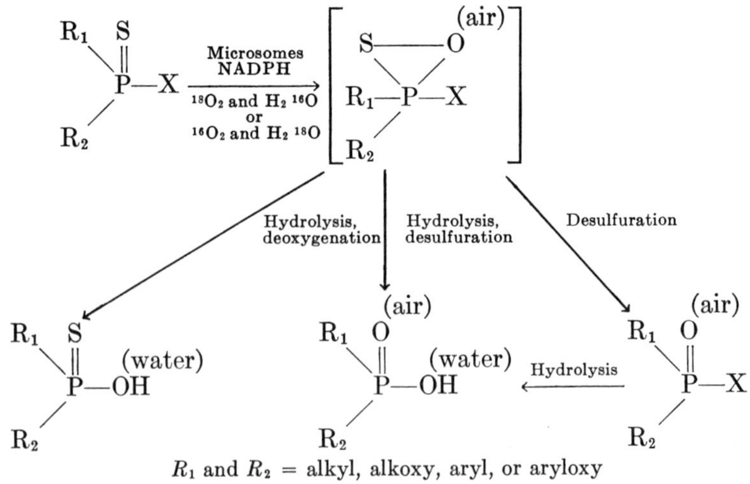

Fig. 6. Proposed mechanism for the NADPH-dependent microsomal oxidase metabolism of thionophosphorus insecticide chemicals (after MCBAIN *et al.* 1971, PTASHNE *et al.* 1971).

tem to the corresponding diethyl phosphorus acids (McBAIN *et al.* 1971 b, NAKATSUGAWA *et al.* 1969), nor do oxons in general serve as good substrates for oxidative cleavage of the anhydride bond to form disubstituted phosphorus acids (NAKATSUGAWA *et al.* 1968).

Most thionophosphorus insecticides and their respective oxons, in addition to the acid anhydride bond, possess one or two alkoxy phosphorus bonds potentially susceptible to cleavage (desalkylation) predominantly via two mechanisms including oxidative cleavage by the NADPH-dependent microsomal oxidase system or cleavage by a reduced glutathione-dependent alkyl transferase system (BULL 1972, DAUTERMAN 1971).

In addition to the major initial cleavage sites already considered, there also occurs with organophosphorus compounds further degradation to monosubstituted phosphorus acids and, in the case of phosphate esters, to inorganic phosphate. Knowledge of the mechanisms that might be involved in these secondary cleavage reactions is with few exceptions largely fragmentary. Mammalian tissue preparations by nonoxidative mechanisms sequentially degraded demethyl dichlorvos, but not dimethyl phosphate, to monomethyl phosphate and inorganic phosphate (HODGSON and CASIDA, 1962). Parathion in the rat is, in addition to other products, metabolized to ethyl phosphoric acid and inorganic phosphate (NAKATSUGAWA *et al.* 1969), but while desethyl paraoxon is extensively metabolized in rats (APPLETON and NAKATSUGAWA 1972), the following products are excreted unchanged: desethyl parathion (NAKATSUGAWA *et al.* 1969), diethyl phosphorothioic acid (NAKATSUGAWA *et al.* 1969, VINOPAL and FUKUTO 1971), and diethyl phosphoric acid (APPLETON and NAKATSUGAWA 1972). In addition, *O*-ethyl ethylphosphonothioic acid is not further biotransformed in the rat (McBAIN *et al.* 1971 a).

Thus, there is some evidence to suggest that the desalkylation products of oxons serve as the substrates for secondary cleavage reactions.

Little is known about the systems involved in the metabolism of organophosphorus pesticides in plants, but usually the same or similar metabolic products are formed in plants or animals; thus, the mechanisms involved may be comparable (CASIDA and LYKKEN 1969, BULL 1972, MENN 1972). Higher plants possess a microsomal oxidase system similar to that found in mammals, and a peroxidase system, both of which have the potential for catalyzing the oxidative transformations observed *in vivo* in plants.

IV. Comparative toxicity of phosphonate and phosphate insecticides

Comparative toxicity studies made with phosphonates and their phosphate analogs (Table III), show that the phosphonates are almost

Table III. *Comparative toxicity of analogous phosphonate-phosphate insecticide pairs.*[a]

Compound	Toxicity (LD_{50})[b]			Cholinesterase inhibition (Molar I_{50}) $\times 10^7$	
	Housefly topical (μg/g)		Mammalian oral[d] (mg/kg)		
	Susceptible	Resistant[c]		Fly head	Mouse brain
I. Methyl parathion	1.2 (0.7)	89 (3.7)	23 (3.7)	—	—
II. Methyl paraoxon	2.5 (1.5)	25 (6.3)	14 (1.0)	1.0 (0.95)	3.3 (0.95)
III. Sumithion	3.1 (1.9)	126 (13.0)	1,250 (14.0)	—	—
IV. Sumioxon	4.3 (2.2)	141 (50.0)	120 (—)	0.56 (0.38)	19.0 (3.4)
V. Paraoxon	1.2 (1.2)	— (—)	— (—)	0.26 (0.13)	—
VI. Imidan	4.0 (10.0)	— (—)	~250 (5.8)	—	—
VII. *O,O*-Diethyl *S*-phenyl phosphorodithioate	56.0 (4.3)	1,600 (40.0)	233 (16.0)	—	—
VIII. *O,O*-Diethyl *S*-4-methylphenyl phosphorodithioate	30.0 (4.3)	1,100 (23.0)	316 (123.0)	—	—

[a] From HOLLINGWORTH *et al.* (1967) (I–IV), FUKUTO and METCALF (1959) (V), SCHRADER (1963) (II), SZABO and MENN (1965 and 1969) (VI–VIII).
[b] Data for the phosphonate analogs of the listed phosphates are enclosed by parentheses.
[c] *S*-Chlorthion-resistant houseflies.
[d] Mice were used with I, III, and IV, and rats with II and VI–VIII.

invariably more toxic to rats, mice, and resistant and susceptible insects. Furthermore, the phosphonate oxons are also better inhibitors of housefly head and mouse brain cholinesterases than their phosphate analogs. In addition to the compounds considered in Table III, acute toxicity data (rat oral LD_{50} in mg/kg) are available for the following commercial phosphates and their phosphonate analogs (SCHRADER 1963), with the LD_{50}'s for the phosphonate analogs shown in parentheses: coumaphos 100 (25), disulfoton 2.6 to 12.5 (2), ethion 55 (10), fenthion 215 to 245 (5), nemacide 270 (75), phorate 1.1 to 2.3 (1), thiometon 85 (25), and carbophenothion 30 (10). Again, the phosphonates are more toxic than their corresponding phosphate analogs.

The observation from metabolism studies that there is a greater accumulation of phosphonate oxons than the corresponding phosphate oxons (Table II) may explain in part why phosphonate insecticides are frequently more acutely toxic to animals than their phosphate analogs.

V. Biological stability of the P—C bond

The biological stability of the P—C bond in higher organisms was first demonstrated in studies which showed that ^{32}P-labeled methylphosphonic and *O*-isopropyl methylphosphonic acids are excreted unchanged in rat urine (HOSKIN 1956) and that α-aminophosphonic acids failed to convert to inorganic phosphate when incubated with liver, kidney, or plant tissue preparations (RYZKOV *et al.* 1954).

Among the examples of alkyl- and phenylphosphonates considered in this review, the P—C bond in most instances remains intact in mammals and higher plants. The most notable exception, as already mentioned, occurs with the 2-chloroethylphosphonates (trichlorfon, Butonate, and ethephon) which readily undergo P—C bond cleavage in biological and nonbiological systems. Another apparent exception involves the phenylphosphonate Inezin which, in rice plants, apparently yields a phosphate metabolite.

Among microorganisms, however, the capacity to catabolize the P—C bond appears to be widespread. A number of bacterial strains have been described as capable of utilizing, as their sole source of phosphorus, the naturally occurring 2-aminoethylphosphonic acid (2-AEP) and various other aminoalkylphosphonate analogs (HARKNESS 1966) and methyl- and ethylphosphonic acids (ZELEZNICK *et al.* 1963). It is implied that bacteria sustained by a phosphonate medium are able to convert the phosphonates to inorganic phosphate. Indeed, only inorganic phosphate was detected in the hydrolysate of a strain of *Escherichia coli* grown on 2-AE^{32}P (HARKNESS 1966).

VI. Conclusions

This review focused attention on the biotransformation and metabolic fate of a group of biologically active phosphonate esters, primarily insecticides. Although many of the comparative aspects of metabolism of phosphates and phosphonates have been elucidated, further work is needed to delineate the reactions in biological and nonbiological systems which govern the lability of the P—C bond.

Summary

The metabolism of a number of alkyl- and phenylphosphonates representing insecticides, a fungicide, and a plant-growth regulator is reviewed with special reference to the fate of the phosphorus moiety, the comparative toxicity of phosphonates and their analogous phosphates to mammals and insects, and the ultimate disposition of the P—C bond of the phosphonate moiety.

In most instances, the metabolism of phosphonates in mammals and higher plants is similar to that so abundantly described for phosphates except that the P—C bond usually remains intact and the terminal metabolites are alkyl- or arylphosphonic acid derivatives. Certain phosphonates are unstable and these include the 2-chloroethylphosphonate insecticides, trichlorfon and Butonate, and the plant-growth regulator, ethephon, which readily undergo P—C bond cleavage in both biological and nonbiological systems. A phenylphosphonate fungicide, Inezin, apparently also cleaves in plants at the P—C linkage.

Many bacterial strains have been described with the ability to catabolize the phosphonate P—C bond. Thus, alkyl- and arylphosphonate terminal metabolites from animals and plants once passed into the environment are likely to be mineralized by microorganisms and reenter the phosphate pool.

Comparative studies between directly analogous phosphonate-phosphate insecticide pairs show that the phosphonates are generally more toxic to mammals and insects and their oxons are better inhibitors of cholinesterase.

The greater toxicity of phosphonates to certain injurious OP-resistant insect species may be the result of more efficient intoxication by the phosphonate oxons.

References

AHMED, M. K., J. E. CASIDA, and R. E. NICHOLS: Bovine metabolism of organophosphorus insecticides: Significance of rumen fluid with particular reference to parathion. J. Agr. Food Chem. **6**, 740 (1958).

APPLETON, H. T., and T. NAKATSUGAWA: Paraoxon deethylation in the metabolism of parathion. Pest. Biochem. Physiol. **2**, 286 (1972).

BULL, D. L.: Metabolism of organophosphorus insecticides in animals and plants. Residue Reviews **43**, 1 (1972).

CASIDA, J. E., and L. LYKKEN: Metabolism of organic pesticide chemicals in higher plants. Ann. Rev. Plant Physiol. **20**, 607 (1969).

DAUTERMAN, W. C.: Biological and nonbiological modifications of organophosphorus compounds. Bull. W.H.O. **44**, 133 (1971).

ENDO, K., Y. MORI, K. KAKIKI, and T. MISATO: Studies on absorption, translocation and metabolic fate of radio active Inezin in rice plant. Nippon Nogei Kagaku Kaishi **44**, 356 (1970).

FUKUTO, T. R., and R. L. METCALF: The effect of structure on the reactivity of alkylphosphonate esters. J. Amer. Chem. Soc. **81**, 372 (1959).

HAMBROOK, J. L., D. J. HOWELLS, and D. UTLEY: Degradation of phosphonates. Breakdown of Soman (*O*-pinacolyl methylphosphonofluoridate) in wheat plants. Pest. Sci. **2**, 172 (1971).

HARKNESS, D. R.: Bacterial growth on aminoalkylphosphonic acids. J. Bacteriol. **92**, 623 (1966).

HARRIS, L. W., L. M. BRASWELL, J. P. FLEISHER, and W. J. CLIFF: Metabolites of pinacolyl methylphosphonofluoridate (Soman) after enzymatic hydrolysis *in vitro*. Biochem. Pharmacol. **13**, 1129 (1964).

HODGSON, E., and J. E. CASIDA: Mammalian enzymes involved in the degradation of 2,2-dichlorovinyl dimethyl phosphate. J. Agr. Food Chem. **10**, 208 (1962).

HOLLINGWORTH, R. M., T. R. FUKUTO, and R. L. METCALF: Selectivity of Sumithion compared with methyl parathion. Influence of structure on anticholinesterase activity. J. Agr. Food Chem. **15**, 235 (1967 a).

——, R. L. METCALF, and T. R. FUKUTO: The selectivity of Sumithion compared with methyl parathion. Metabolism in the white mouse. J. Agr. Food Chem. **15**, 242 (1967 b).

HOLMSTEAD, R. L., T. R. FUKUTO, and R. B. MARCH: The metabolism of *O*-(4-bromo-2,5-dichlorophenyl) *O*-methyl phenylphosphonothioate (leptophos) in the white mouse and on the cotton plant. Arch. Environ. Contam. Toxicol. **1**, 133 (1973).

HOSKIN, F. C. G.: Some observations concerning the biochemical inertness of methylphosphonic and isopropyl methylphosphonic acids. Canad. J. Biochem. **34**, 743 (1956).

KENAGA, E. E., and W. E. ALLISON: Commercial and experimental organic insecticides (1971 Revision). Bull. Entomol. Soc. Amer. **15**, 85 (1969).

MARCO, G. J., and E. G. JAWORSKI: Metabolism of *O*-phenyl-*O'*-(4-nitrophenyl) methylphosphonothionate (Colep) in plants and animals. J. Agr. Food Chem. **12**, 305 (1964).

MCBAIN, J. B., and J. J. MENN: Spontaneous conversion of alkylphosphonodithioate insecticides to cholinesterase inhibitors. Presented Ann. Meeting Entomol. Soc. Amer., Philadelphia, Pa., Nov. 30-Dec. 3 (1964).

——, L. J. HOFFMAN, and J. J. MENN: Metabolic degradation of *O*-ethyl S-phenyl ethylphosphonodithioate (Dyfonate) in potato plants. J. Agr. Food Chem. **18**, 1139 (1970).

—— —— —— Dyfonate metabolism studies. II. Metabolic pathway of *O*-ethyl S-phenyl ethylphosphonodithioate in rats. Pest. Biochem. Physiol. **1**, 356 (1971 a).

——, I. YAMAMOTO, and J. E. CASIDA: Mechanism of activation and deactivation of Dyfonate (*O*-ethyl S-phenyl ethylphosphonodithioate) by rat liver microsomes. Life Sci. **10** (Part II), 947 (1971 b).

MENN, J. J.: Terminal residues of phosphonate insecticides. In: Pesticide terminal residues, Tel Aviv, Israel (1971); Suppl. Pure Applied Chem., p. 57 (1971).

—— Absorption and metabolism of insecticide chemicals in plants. In: Degradation of synthetic chemicals in the biosphere. Washington, D.C.: Nat. Acad. Sci. (1972).

——, and K. SZABO: The synthesis and biological properties of new *O*-alkyl S-aryl alkylphosphonodithioates. J. Econ. Entomol. **58**, 734 (1965).

NAKATSUGAWA, T., N. M. TOLMAN, and P. A. DAHM: Degradation and activation of parathion analogs by microsomal enzymes. Biochem. Pharmacol. **17**, 1517 (1968).

—— —— —— Degradation of parathion in the rat. Biochem. Pharmacol. **18**, 1103 (1969).

PTASHNE, K. A., R. M. WOLCOT, and R. A. NEAL: Oxygen-18 studies on the chemical mechanisms of the mixed function oxidase catalyzed desulfuration and dearylation reactions of parathion. J. Pharmacol. Expt. Therap. **179**, 380 (1971).

RYZKOV, V. L., M. I. KABACHNIK, L. M. TARASEVICH, T. YA. MEDVED, N. A. ZEĬTLENOK, N. K. MARCHENKO, V. A. VAGZHANOVA, E. F. ULANOVA, N. V. CHEBURKINA, and D. I. IVANOVSKIĬ: Biological activity of α-aminophosphonic acids. Doklady Akad. Nauk S.S.S.R. **98**, 849 (1954); through Chem. Abstr. **49**, 3404 (1955).

SCHRADER, G.: Die Entwicklung neuer insektizider Phosphorsäure-Ester. 3., neubearbeitete Aufl. Weinheim/Bergstr: Verlag Chemie, (1963).

—— Synthesis and properties of important organophosphorus insecticides. Pflanzensch. Ber. **36**, 29 (1967).

SZABO, K., and J. J. MENN: Synthesis and biological properties of insecticidal *N*-(mercaptomethyl) phthalimide-S-(*O*-alkyl)-alkylphosphonodithioates and thiolates. J. Agr. Food Chem. **17**, 863 (1969).

YAMAGUCHI, M., C. W. CHU, and S. F. YANG: The fate of ^{14}C (2-chloroethyl) phosphonic acid in summer squash, cucumber, and tomato. J. Amer. Soc. Hort. Sci. **96**, 606 (1971).

YANG, S. F.: Ethylene evolution from 2-chloroethylphosphonic acid. Plant Physiol. **44**, 1203 (1969).

ZELEZNICK, L. D., T. C. MYERS, and E. B. TITCHENER: Growth of *Escherichia coli* on methyl- and ethylphosphonic acids. Biochem. Biophys. Acta **78**, 546 (1963).

Manuscript received September 4, 1973; accepted October 17, 1973.

New aspects of organophosphorus pesticides. V. Oxidative rearrangement of organosphosphorus thionate esters.

By

D. A. WUSTNER,* M. A. FAHMY,* and T. R. FUKUTO*

Contents

I. Introduction

Oxidative pathways account for much of the *in vivo* chemical modification of pesticides, both activation and detoxication. In fact, the oxidative desulfurization of phosphorothionates, such as parathion, to the corresponding phosphate is essential for the toxicity of organophosphorothionate insecticides. A wide variety of other reactions is catalyzed by this system including thioether oxidation, *N*- and *O*-dealkylation, aromatic hydroxylation, and epoxidation. In living systems, the enzymes responsible for these oxidative reactions are located in the smooth endoplasmic reticula and are called the multifunction oxidase (MFO) system. A preparation of endoplasmic reticula isolated by cellular homogenation and differential centrifugation is called a microsomal system. For oxidative activity the *in vitro* microsomal preparations must be fortified with NADPH and molecular oxygen.

Because of the biological importance of these multi-function oxidase systems, a number of chemical model oxidative systems have

* Division of Toxicology and Physiology, Department of Entomology, University of California, Riverside, California 92502.

This work was supported by Research Grant No. R801837 from the Environmental Protection Agency, Washington, D.C., a Research-Training Grant from The Rockefeller Foundation, New York, N.Y., and Research Grant 73-261 from Cotton Incorporated, Raleigh, N.C.

been used. These include Fenton's reagent of hydrogen peroxide and ferrous salts (FENTON 1894), the Udenfriend system of ascorbic acid, ferrous sulfate, and oxygen (UDENFRIEND *et al.* 1954), peroxytrifluoro acetic acid, and *m*-chloroperoxybenzoic acid (HERRIOTT 1971). The major justification for the use of these systems in all their variety is that the mechanism(s) for biological MFO is not well understood and that, in many cases, the products of the chemical systems are similar to those of the biological systems. Thus, subjecting a compound to one of these model systems has considerable predictive value as to what the *in vivo* metabolites are likely to be. In addition, it is hoped that a study of these model systems will lead to a better understanding of biological MFO.

The topic that will be discussed is an oxidative rearrangement reaction that is observed when phosphorothionate (P=S) compounds are treated with *m*-chloroperoxybenzoic acid in dichloromethane. Additionally, the possible significance of this rearrangement to the products from microsomal systems will be briefly discussed.

II. Oxidation of Dyfonate®

Treatment of Dyfonate® (*O*-ethyl *S*-phenyl ethylphosphonodithioate) (*1*) with *m*-chloroperoxybenzoic acid in dichloromethane yields, as shown in Figure 1, seven identifiable products (MCBAIN *et al.* 1971 b). All but one of these products, unreacted Dyfonate® (*1*), Dyfonate oxon (*2*), *O*-ethyl ethylphosphonic acid (*3*), *O*-ethyl ethylphosphonothioic acid (*4*), sulfur, and diphenyl disulfide, were expected and easily identifiable. The oxidatively rearranged phenyl ethoxy(ethyl)phosphinyl disulfide (Dyfonate oxon disulfide) (*5*), an oxidatively rearranged product, was not expected.

Dyfonate oxon disulfide (*5*) may be synthesized by treating *O*-ethyl ethylphosphonothioic acid in carbon tetrachloride with benzene sulfenyl chloride by the procedure of SCHRADER and LORENZ (1959), below. The product is purified on a silica gel column eluted with ether (WUSTNER *et al.* 1972).

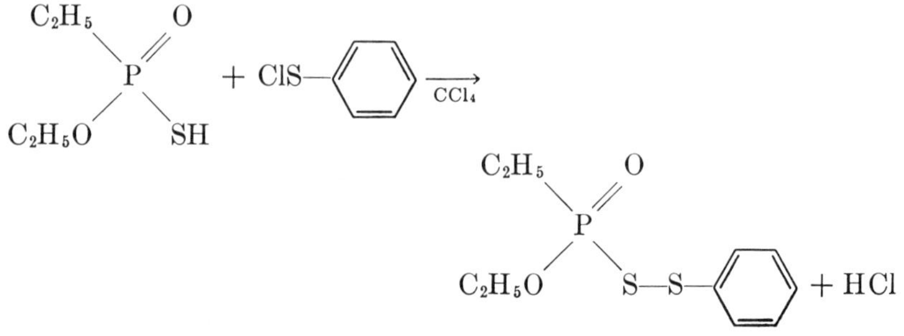

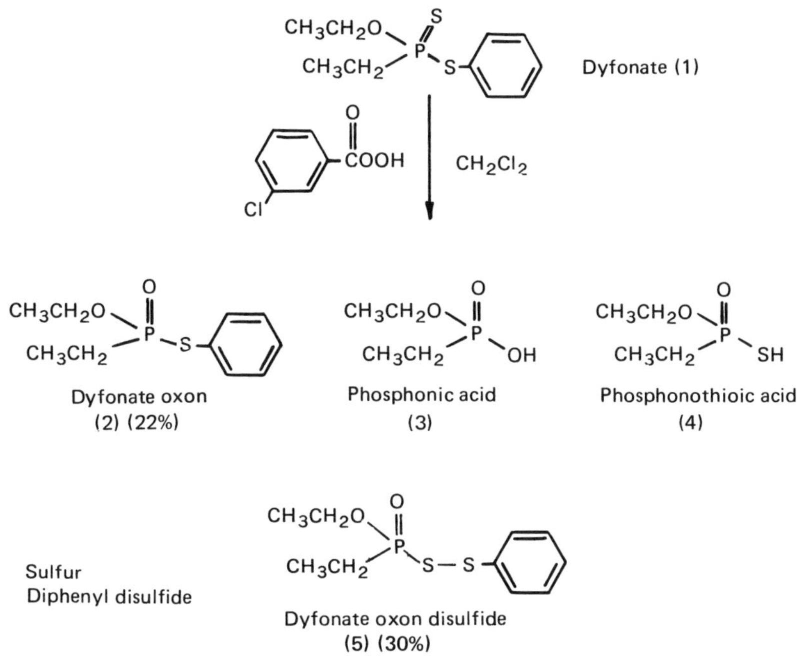

Fig. 1. The reaction of Dyfonate® with *m*-chloroperoxybenzoic acid in dichloromethane.

The proton NMR and infrared spectra of Dyfonate oxon (2) and Dyfonate oxon disulfide (5) are nearly identical. For example, the infrared spectra of Dyfonate®, Dyfonate oxon, and Dyfonate oxon disulfide are shown in Figure 2. Elemental analysis and molecular weight determination revealed that the compound was similar to Dyfonate® with the addition of a single atom of oxygen. The disulfide showed a unique 31P NMR signal (McBain *et al.* 1971 b).

The only unequivocal spectral evidence for the structure of Dyfonate oxon disulfide (5) comes from mass spectrometry (Fig. 3). A molecular ion at m/e 262 is observed just 16 higher than Dyfonate® and 32 higher than Dyfonate oxon. More importantly, the fragmentation pattern identifies the disulfide. Fragments are seen that correspond to the loss of phenyl-S and phenyl-S_2 (m/e 121 and 154) and fragments were detected that correspond to phenyl-S and phenyl-S_2 (m/e 109 and 141). The base peak at m/e 93 corresponds to $C_2H_6PO_2$ which is a common fragment in the mass spectra of ethylphosphonates. The relative intensities observed for the mass spectral fragments are the same whether the Dyfonate oxon disulfide is ob-

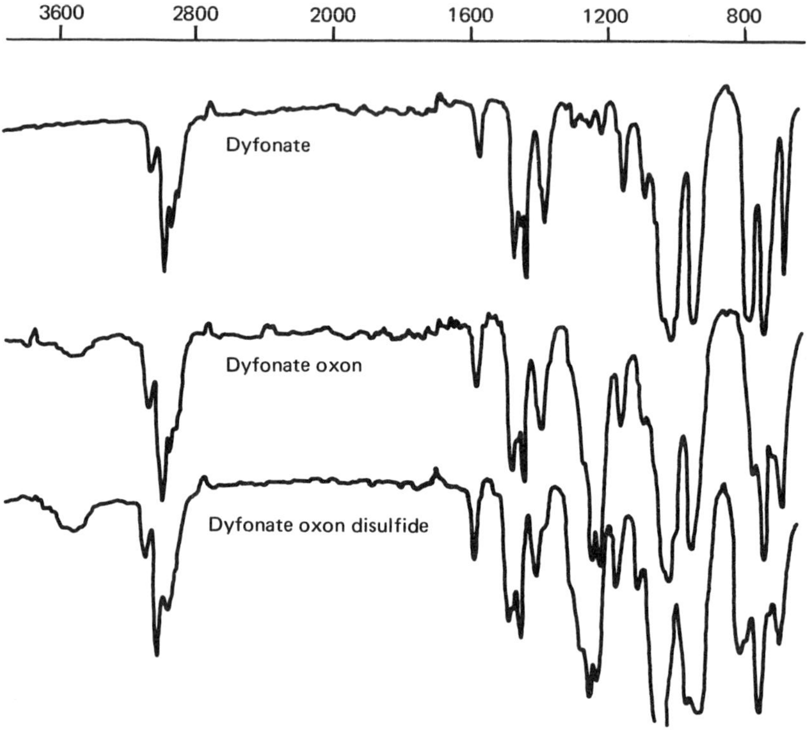

Fig. 2. Infrared spectra of Dyfonate® (*1*), Dyfonate oxon (*2*), and Dyfonate oxon disulfide (*5*); the spectra of *2* and *5* are almost identical.

tained from the *m*-chloroperoxybenzoic acid oxidation of Dyfonate® or from direct synthesis (Table I).

When Dyfonate oxon disulfide is exposed to the atmosphere on silica gel plates or is treated with water, the decomposition products are the two phosphonic acids with the thioic acid (*4*) predominating

Table I. *Relative intensities of mass spectral fragments from Dyfonate oxon disulfide from both peracid oxidation of Dyfonate® and independent synthesis.*

m/e	65	77	78	93	97	109	110
Peracid	36.0	12.4	15.2	100	6	22.4	12.1
Synthetic	35.1	12.0	14.0	100	5.4	21.6	10.8

m/e	121	141	142	154	262	263	264
Peracid	36.4	9.2	4.4	4.8	19.1	2.8	2.0
Synthetic	36.0	8.3	3.5	4.4	18.1	2.7	2.0

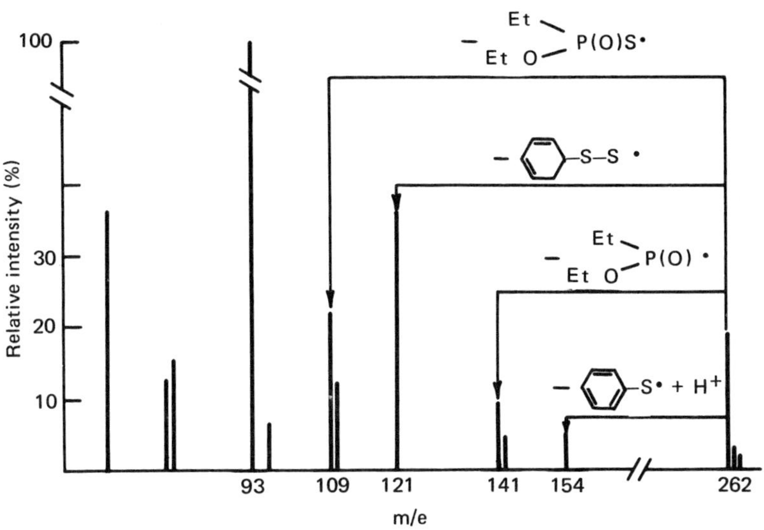

Fig. 3. Mass spectrum of Dyfonate oxon disulfide (5) showing those fragmentations that particularly support its disulfide structure.

and diphenyl disulfide:

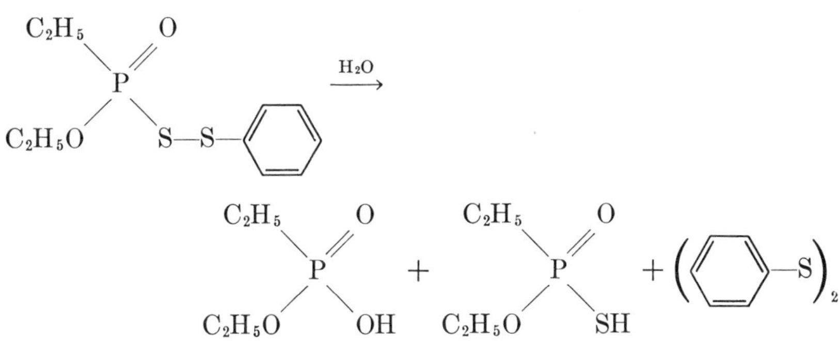

The mechanism proposed for this oxidative rearrangement is similar to the mechanism proposed for the next oxidative rearrangement reaction that will be discussed and so both will be presented together.

III. Oxidation of *N*-(dimethoxyphosphinothioyl)-carbofuran

One of our primary research interests is in insecticide selectivity, *i.e.*, insecticides that are toxic to insects and substantially less toxic to mammals. Shown below is the structure of the carbamate insecticide carbofuran (9). Unfortunately, it is not a selective

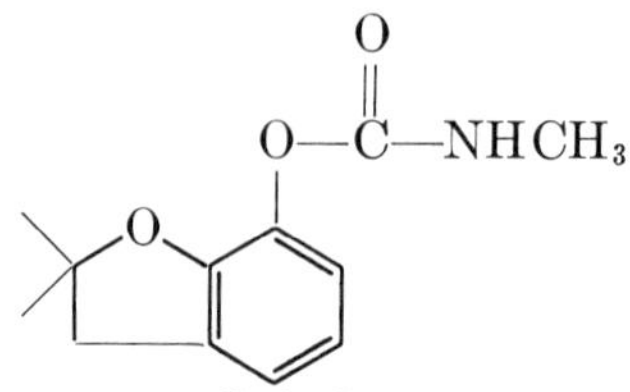

Carbofuran

insecticide, for carbofuran is highly toxic to both insects and mammals. We have found that substitution of the *N*-hydrogen with a variety of groups provides *N*-methylcarbamates with favorable properties of selectivity. 2,2-Dimethyl-2,3-dihydrobenzofuranyl-7 *N*-methyl-(dimethoxyphosphinothioyl)carbamate (*N*-[dimethoxyphosphinothioyl]-carbofuran (6) is such a compound.

When *N*-(phosphinothioyl)-carbofuran (6) is treated with *m*-chloroperoxybenzoic acid in dichloromethane (Fig. 4) a rearrange-

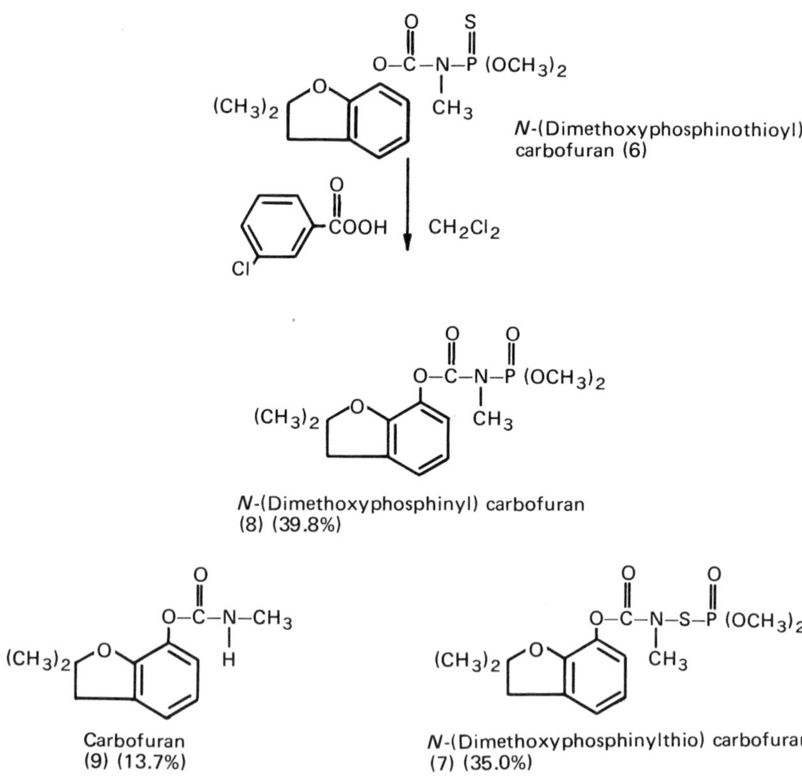

Fig. 4. The reaction of *N*-(dimethoxyphosphinothioyl)-carbofuran (6) with *m*-chloroperoxybenzoic acid in dichloromethane.

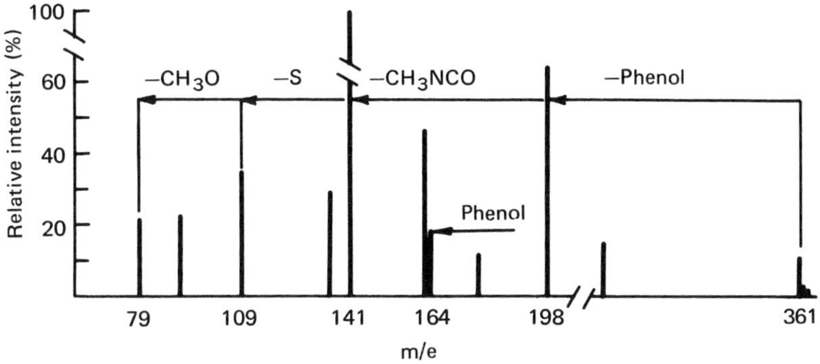

Fig. 5. Mass spectrum of *N*-(dimethoxyphosphinothiol)-carbofuran (*7*).

ment reaction similar to that previously described for Dyfonate® occurs to give the rearranged *N*-(dimethoxyphosphinothio)-carbofuran (*7*) in addition to the expected *N*-(dimethoxyphosphinyl)-carbofuran (oxon, *8*) as the major products along with lesser amounts of carbofuran (*9*) and starting material (FAHMY and FUKUTO 1972). Sulfur was also isolated. Quantitation of this reaction was based on a radioactive sample with 95 percent recovery.

Attempts to synthesize directly either the rearranged product (*7*) or the oxon (*8*) failed and so proof of structure must rest on spectral, chemical, and analytical data.

Figure 5 shows the mass spectrum of the oxidatively rearranged *N*-(dimethoxyphosphinyl)-carbofuran (*7*). The fragmentations that are labeled are those that particularly support the assigned structure. The parent ion (m/e 361) loses the phenol to give the *N*-(dimethoxyphosphinylthiol)carbamoyl moiety (m/e 198). This fragment then loses methylisocyanate to give dimethyl phosphorothioate (m/e 141). Loss of sulfur and then formaldehyde gives peaks at m/e 109 and 79. The other compounds, *i.e.*, the starting material (*6*) and the oxon (*8*), give analogous fragmentation patterns, *i.e.*, loss of phenol, then methylisocyanate, and finally typical organophosphorus fragmentations.

Further support of the oxidatively rearranged compound (*7*) was obtained from its decomposition on silica gel thin-layer plates as seen in Figure 6. Quantitation was again accomplished by a radioactive sample with 97 percent total recovery. The structure of the *N,N'*-bis-carbofuran disulfide (*10*) was proved by spectral evidence and elemental analysis. In this experiment, a small amount of the oxon (*8*) was detected. Because the $\mathbf{R}_f$ values of compounds *7* and *8* are rather close in the systems used, it is probable that compound *8* was a contamination in the sample of compound *7*. Attempts to synthesize the disulfide (*10*) led to the discovery of the selective properties of *N,N'*-bis-dicarbamoyl sulfides (FAHMY *et al.* 1973).

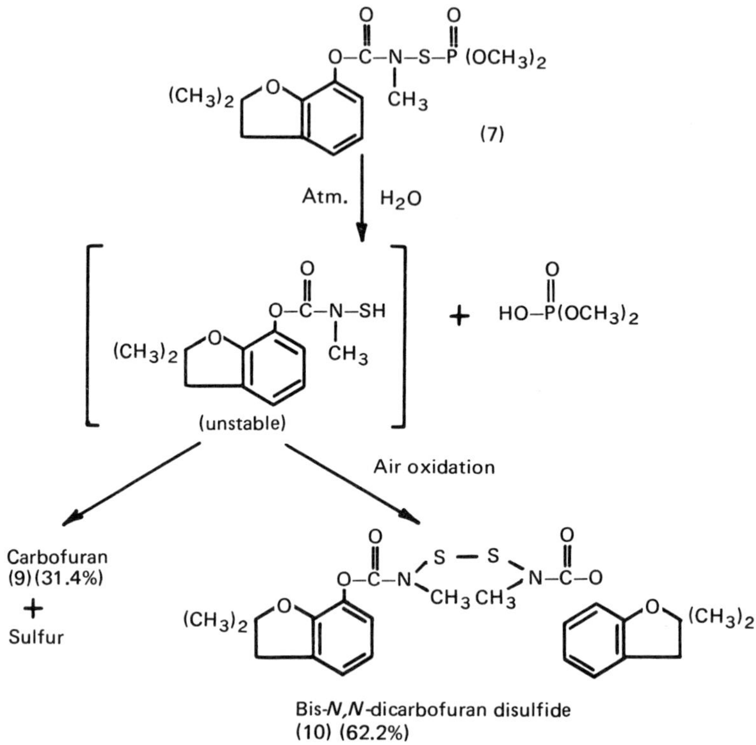

Fig. 6. Decomposition of N-(dimethoxyphosphinothiol)-carbofuran (7) on silica gel TLC plates.

IV. Mechanisms

The products of both the Dyfonate® and N-(dimethoxyphosphinothioyl)-carbofuran oxidations can be explained by assuming that an unstable S-oxide intermediate is initially formed. This type of intermediate has been used by Herriott (1971) to account for the stereochemistry of the oxidation of phosphinothioates and by Nilsson and Senning (1972) to explain the oxidative rearrangement of C-sulfonylthioformamidines. This S-oxide intermediate may be written in a number of resonance and tautomeric forms:

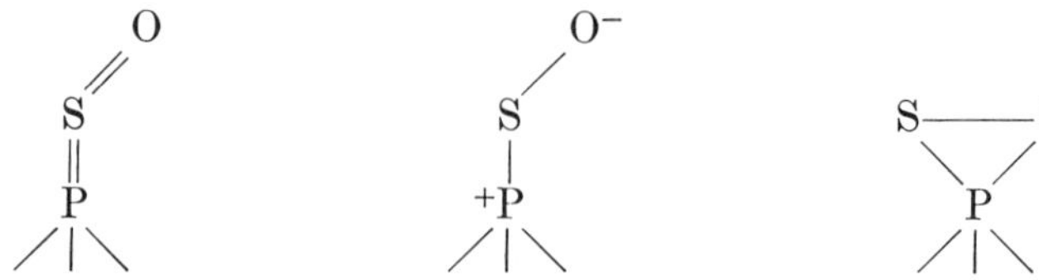

The formation of the corresponding oxon and rearranged product is easily rationalized from the 3-membered ring tautomer. Shown in Figure 7 is a mechanism for the Dyfonate® oxidation. A similar mechanism may be written for the *N*-(dimethoxyphosphinothioyl)-carbofuran oxidation (Fig. 8).

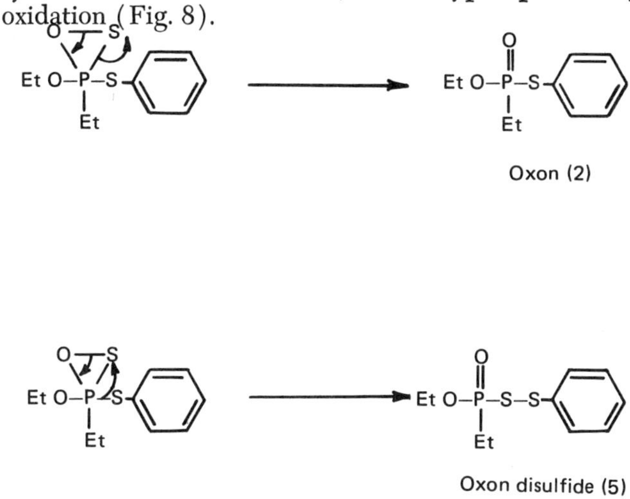

Fig. 7. Proposed mechanism for the formation of Dyfonate oxon (*2*) and Dyfonate oxon disulfide (*5*) from an S-oxide intermediate from Dyfonate®.

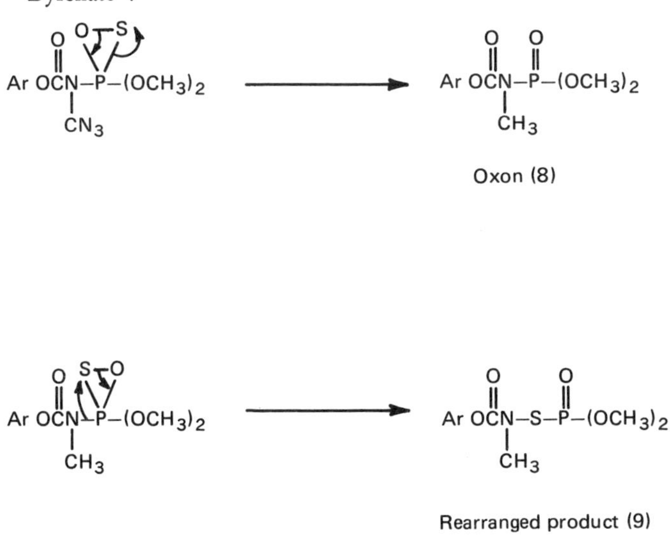

Fig. 8. A mechanism for the the formation of the oxon (*8*) and the rearranged product (*9*) from an S-oxide intermediate from *N*-(dimethoxyphosphinothioyl)-carbofuran; this proposed mechanism is similar to that shown in Figure 7 for Dyfonate®.

Table II. *Incorporation of oxygen from molecular oxygen and water into the products of microsomal oxidation of parathion and Dyfonate®.*

Compound	Reaction	Product	% ^{18}O from	
			$H_2{}^{18}O$	$^{18}O_2$
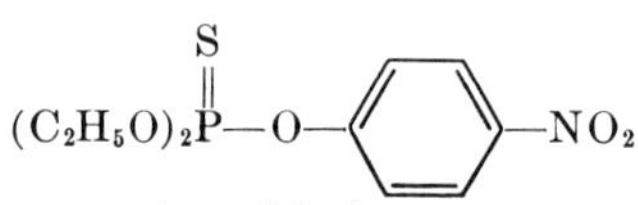 (parathion)		oxon	0	100
	Microsomes → NADPH O_2	thioic acid	100	0
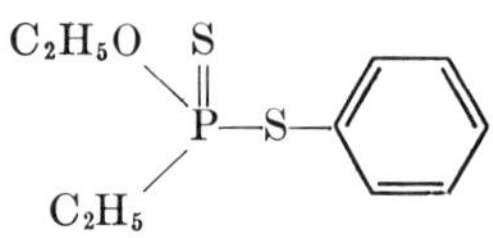 (Dyfonate®)		oxygen acid	50	50

The Dyfonate oxon disulfide (5) has been tentatively reported as a product from the *in vitro* microsomal oxidation of Dyfonate® (McBAIN *et al.* 1971 b). Experiments have been made to determine the source of the oxygen incorporated into the products from the microsomal oxidation of parathion (PTASHNE *et al.* 1971) and Dyfonate® (McBAIN *et al.* 1971 a), and these results are shown in Table II.

In these *in vitro* yet biological systems all of the oxygen incorporated into the oxons comes from molecular $^{18}O_2$. Conversely, all of the oxygen in the thioic acids comes from water $H_2{}^{18}O$. The oxygens in the phosphoric acids come 500 percent each from both molecular oxygen and water. In both sets of experiments controls were used to assure that the acids were not produced by nonoxidative hydrolytic reactions.

What sort of mechanism can be invoked for microsomal oxidation to explain both a possible rearranged intermediate and the oxygen incorporation from the ^{18}O experiments? An attractive possibility that can be written for these reactions is shown in Figures 9 and 10.

The incorporation of atmospheric O_2 into the oxons is explained by the S-oxide intermediate losing elemental sulfur. From the same intermediate, rearrangement in the two possible directions and subsequent attack by water on phosphorus yields the appropriately labeled acids.

The decomposition of the *N*-(dimethoxyphosphinylthio)-carbofuran (7) discussed previously appears to follow this pathway. The initial decomposition yields dimethylphosphoric acid and apparently an unstable carbofuran-*N*-thiol. The thiol then decomposes to the

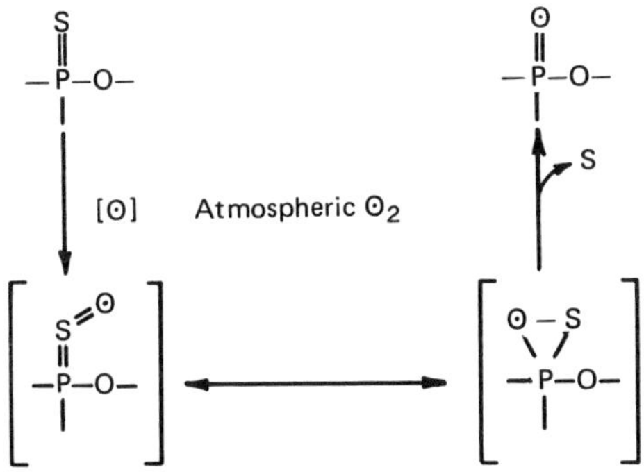

Fig. 9. Proposed S-oxide mechanism for the incorporation of molecular oxygen by microsomal systems into oxons.

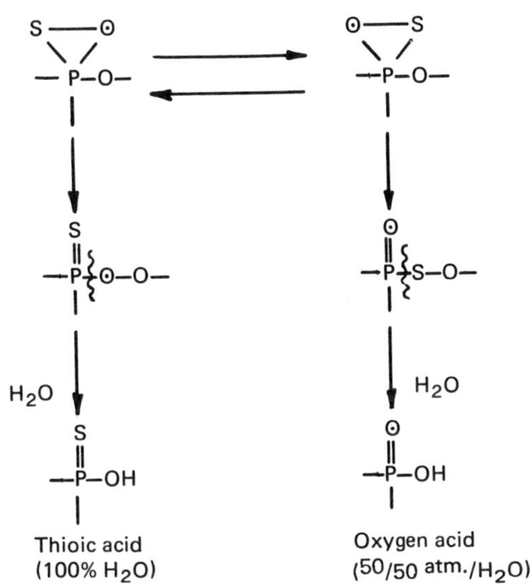

Fig. 10. Proposed S-oxide mechanism for the incorporation of molecular oxygen and oxygen from water into the acids formed from microsomal systems; as in Figure 9, oxygen from O_2 is marked with a central dot.

carbamate plus sulfur or is oxidized to the *N,N′*-bis-dicarbofuran disulfide (*10*). However, the Dyfonate oxon disulfide (*5*) decomposes in water to both the phosphonothioic acid (*4*) by S—S bond cleavage and the expected phosphonic acid (*3*) by P—S bond cleavage. Disulfide bond cleavage is not consistent with water attack on phosphorus, and is thus not consistent with the mechanism proposed. It is conceivable that the phosphorus in an enzymatically bound rearranged intermediate or product might be preferentially attacked by water.

However, invoking special mechanisms for enzymatic reactions to rationalize inconsistent chemical data is not a satisfactory solution. Thus, it is probable or possible that oxidative rearrangement products such as Dyfonate oxon disulfide are not significant intermediates in biological oxidations of P=S to P=O.

Work is continuing in our laboratories on this reaction including an investigation of the stereochemistry of the rearrangement using optically active substrates of known absolute configurations.

Summary

Biological oxidations are of great importance in the *in vivo* modification of pesticides. A number of chemical oxidation systems have been proposed as models of these biological processes. Treatment of the organophosphorus thionate esters Dyfonate® and *N*-(dimethoxyphosphinothioyl)-carbofuran with *m*-chloroperoxybenzoic acid in dichloromethane yields both the expected oxons as well as rearrangement products. A mechanism is proposed that assumes that an initial S-oxide intermediate is formed that then either extrudes sulfur to form the oxon or rearranges to form, from Dyfonate®, phenyl ethoxy(ethyl)phosphinyl disulfide. This disulfide decomposes into both *O*-ethyl ethylphosphonic acid and *O*-ethyl ethylphosphonothioic acid plus diphenyl disulfide. This mode of decomposition is not consistent with ^{18}O labeling data from biological microsomal systems and thus the rearrangement products from the peracid system are probably not significant products in microsomal oxidations.

References

FAHMY, M. A. H., Y. C. CHIU, and T. R. FUKUTO: Selective toxicity of *N*-substituted *bis*-carbamoyl sulfides. J. Agr. Food Chem. **22**, In press (1974).

——, and T. R. FUKUTO: Oxidative rearrangement of *N*-(dimethoxyphosphinothioyl)carbamate esters. Tetrahedron Let. **41**, 4245 (1972).

FENTON, H. S. H.: Oxidation of partaric acid in presence of iron. J. Chem. Soc. **65**, 899 (1894).

HERRIOTT, A. W.: Peroxy acid oxidation of phosphinothioates, a reversal of stereochemistry. J. Amer. Chem. Soc. **93**, 3304 (1971)...

MCBAIN, J. B., I. YAMAMOTO, and J. E. CASIDA: Mechanism of activation and deactivation of Dyfonate® (*O*-ethyl S-phenyl ethylphosphonodithioate) by rat liver microsomes. Life Sci. **10** (II), 947 (1971 a).

—— —— —— Oxygenated intermediate in peracid and microsomal oxidations of the organophosphonothionate insecticide Dyfonate®. Life Sci. **10** (II), 1311 (1971 b).

NILSSON, N. H., and A. SENNING: Oxidative rearrangement of *C*-sulfonylthioformamides. Angew. Chem. Internat. ed. **11**, 295 (1972).

PTASHNE, K. A., R. M. WOLCOTT, and R H. NEAL: Oxygen-18 studies on the chemical mechanisms of the mixed function oxidase catalyzed desulfuration and dearylation reactions of parathion. J. Pharmacol. Exp. Therap. **179**, 380 (1971).

SCHRADER, G., and W. LORENZ: Sulfenylthiophosphate and -thiophosphonate esters. Ger. Pat. 1,103,324 (Oct. 20, 1959); through Chem. Abstr. **56**, 5888c (1962).

UDENFRIEND, S., C. T. CLARKE, J. AXELROD, and B. B. BRODIE: Ascorbic acid in aromatic hydroxylation. I. A model system for aromatic hydroxylation. J. Biol. Chem. **208**, 731 (1954).

WUSTNER, D. A., J. DESMARCHELIER, and T. R. FUKUTO: Structure for the oxygenated product of peracid oxidation of Dyfonate® insecticide (*O*-ethyl S-phenyl ethylphosphonodithioate). Life Sci. **11**(II), 583 (1972).

Manuscript received September 4, 1973; accepted October 17, 1973.

New aspects of organophosphorus pesticides. VI. Organophosphorus insect chemosterilants.

By

ALEXEJ B. BOŘKOVEC*

Contents

I. Introduction

Sexual sterilization of insects, and for that matter of any sexually reproducing organisms, is one of the most promising approaches to a preventive control and management of populations (KNIPLING 1959 and 1966). Since the reproductive potential of a population decreases when the proportion of its sterile members increases, any process that will increase the ratio between the sterile and fertile organisms will reduce the population increment. Ultimately, when the ratio becomes very large, the population will cease to be self-sustaining and will die out.

Sterility, full or partial, can be achieved by treating the organisms with chemicals (BOŘKOVEC 1962 and 1966), and some of the most effective chemosterilants active in invertebrates, particularly in insects, contain phosphorus. In distinction to human birth control pills, most insect chemosterilants are not hormonal antagonists but they are cytostatic or mutagenic agents with a selective or at least preferential activity in the gonads or specifically in the insect's gametes. Ideally, the effects of insects chemosterilants should be irreversible and they should not decrease longevity, mating activity, or the insect's general competitiveness in nature. Although only a few compounds meet all these requirements, their activity implies a direct interaction of the

* Insect Chemosterilants Laboratory, Agricultural Environmental Quality Institute, ARS, U.S. Department of Agriculture, Beltsville, Maryland 20705.

sterilant with cellular components which is a fundamentally different mechanism from that thought to be responsible for the steriod-induced infertility in man. The search for selective cytostatic agents received its greatest impetus and financial support in the beginnings of cancer chemotherapy and many of the earliest chemosterilants were originally synthesized as potential antineoplastic drugs. In broad outlines the relationship between these two types of agents still remains valid but detailed structure-activity relationships are only seldom comparable.

II. Chemistry and biological activity

From a structural standpoint, organophosphorus chemosterilants are all derivatives of phosphoric and phosphorothioic acid or of cyclic azaphosphorines (Fig. 1).

This rather simple classification, however, may be misleading in its implication that, for example, all sterilizing derivatives of phosphoric acid have a similar mode of action in inducing sterility in insects. Therefore, it is appropriate to stress right at the onset of this discussion that the PO, PS, or PN moieties in chemosterilants are only advantageous carriers of the biologically active substitutents replacing the hydroxyl groups in phosphoric acid or hydrogen in the triazatriphosphorine. A more suitable classification should reflect the structural or chemical properties of that substituent.

a) Alkylating organophosphorus chemosterilants

Compounds that are the most versatile in their sterilizing effects are the derivatives of aziridine in which the ring nitrogen is attached to a PO, PS, or PN moiety. As Figures 2–4 show, a considerable variation in the number and type of the alkylating and nonalkylating substituents is possible and the resulting compounds have widely different physical properties (BOŘKOVEC *et al.* 1966, BEROZA and BOŘKOVEC 1964). However, their chemical properties also vary and so do their physiological effects in insects. As a rule, monoaziridines

X = O phosphoric acid
X = S phosphorothioic acid

1,3,5,2,4,6-triazatriphosphorine

Fig. 1. Parent compounds of organophosphorus insect chemosterilants.

are much less effective sterilants than the di- and polyaziridines; however, there is little or no difference between the physiological effects of di- and triaziridines, providing the third substituent in the diaziridines does not become too large. Although the diaziridines (Fig. 3) appear to be exceptionally well suited for structure-activity comparisons no satisfactory explanation can be offered for the often substantial effects exerted by the nonalkylating substituents. Quantitative studies conducted in our laboratory (CHANG *et al.* 1970) with the male housefly, *Musca domestica* L., revealed no consistent differences between the activity of phosphine oxides and sulfides (Table I). The somewhat more consistent differences in both series owing to the

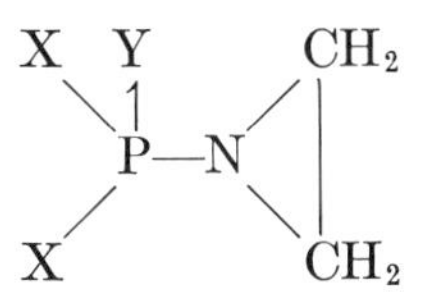

X = OR, NR_2 Y = O, S

Fig. 2. Alkylating phosphorus chemosterilants containing one aziridine group.

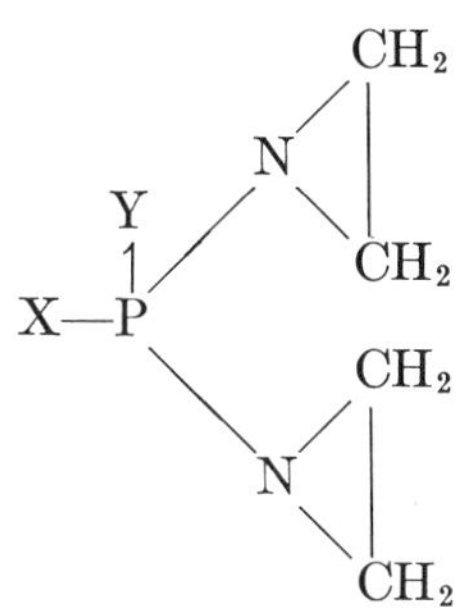

X = R, OR, SR, NR_2
Y = O, S

Fig. 3. Alkylating phosphorus chemosterilants containing two aziridine groups.

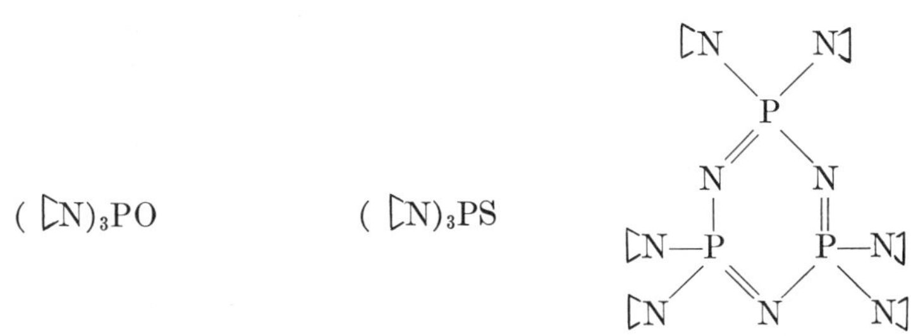

Tepa Thiotepa Apholate

Fig. 4. Alkylating phosphorus chemosterilants containing three or more aziridine groups.

Table I. *Sterilizing effects* (SD_{50}) *of substituted bis(1-aziridinyl)-phosphine oxides and sulfides injected into male houseflies.*

X	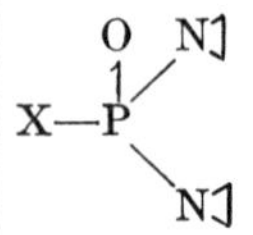	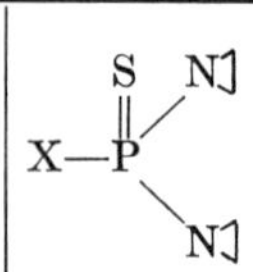
	SD_{50} (nanomoles)	SD_{50} (nanomoles)
—N]	0.46	0.53
—NEt	1.03	.63
—SEt	3.07	1.97
—OEt	3.41	3.85
—Et	4.00	3.35

third substituent on the PO or PS moieties cannot be clearly interpreted in terms of the electronic effects of the substituent. In a series of alkylamino compounds (Table II) the activity decreased with the increase of the carbon chain in the alkyl group (CHANG and BOŘKOVEC 1966). Also, it must be kept in mind that the structure-activity relationships are frequently different in various species and even in the two sexes of the same species. Unfortunately, no quantitative studies that would indicate the reactivity, *i.e.*, the alkylating ability, of these compounds are available but our experience with similar hydrolytical stability of some 40 derivatives of structures shown in Figure 3 does not support the assumption that their *in vitro* activity could be used for predicting their biological behavior. Physiological considerations, however, may at least partly explain the apparent in-

Table II. *Sterilizing* (SD_{50}) *and toxic* (LD_{50}) *effects of bis(1-aziridinyl)-alkylphosphinic amides injected into male houseflies.*

([N)$_2$PO—R R	SD_{50} (nanomoles)	LD_{50} (nanomoles)
$NHCH_3$	0.348	153
$N(CH_3)_2$	1.49	432
NHC_2H_5	1.40	325
$NH(CH_2)_2CH_3$	3.18	11.3
$NHCH(CH_3)_2$	1.21	450
$NH(CH_2)_3CH_3$	4.59	94.3
$NH(CH_2)_7CH_3$	19.1	38.1

consistencies. Since alkylating agents can react with a variety of nucleophilic moieties, they have a wide choice in attacking various cellular constituents. Even at nontoxic but sterilizing doses they frequently attack the cells of the reproductive organs as well as the gametes. At the same time, hydrolytical and other metabolic processes in the hemolymph of the insect inactivate a certain proportion of the alkylating units and thus reduce the effective concentration of the agent before it can enter through the cell wall. In spite of the significance and long history of the physiological activity of the aziridinyl compounds, the question of the ultimate targets that are responsible for lethal or nonlethal mutations is still unanswered. Nucleic acids that are the most commonly suspected targets do not react readily with aziridines but if we realize that, at least in theory, only a few molecules of the sterilant could inactivate one chromosome and thus the entire gamete, the difficulties of the mode of action research become more understandable. Nevertheless, strong evidence for a direct interaction between a gamete and an alkylating agent was presented by TABER and BOŘKOVEC (1969) who treated *in vitro* sperm of male honey bees, *Apis mellifera* L., with different amounts of tepa. Queen bees artificially inseminated with the treated sperm laid sterile eggs and the extent of sterility was proportional to the concentration of tepa.

For obvious reason, chemosterilants must be relatively nontoxic at effective levels. The phosphorus chemosterilants have therefore quite different toxicological properties from those of the related phosphorus insecticides. Most conspicuous is the apparent lack of cholinesterase inhibition exemplified by the relatively low toxicity of the diethyl ester of 1-aziridinylphosphonic acid (Fig. 2, X=OEt and Y=O) that was a sterilant but not a toxicant when added to the diet of houseflies at concentrations as high as five percent (BOŘKOVEC *et al.* 1968). The most thoroughly investigated compound, however, is tepa (Fig. 4) and its toxicology shows all the symptoms of biological alkylating agent and none of the organophosphorus insecticides. Our recent toxicological study in rats showed that at 0.06 mg/kg/day tepa had no apparent toxicological or reproductive effects after 160 days of continuous administration.

The metabolism of alkylating organophosphorus chemosterilants is complex. Fortunately, the opening of the aziridine ring and the subsequent loss of alkylating ability leads invariably to a loss or a drastic decrease of toxicity and sterilizing effects and, therefore, the monitoring of the concentration of the alkylating functions is adequate for most purposes. Though our early work showed that in phosphate-buffered solutions the P—N bond in tepa and in analogous compounds is easily hydrolyzed with a consequent formation of free aziridine (BEROZA and BOŘKOVEC 1964), this process appears to be of minor importance under biological conditions when the aziridinyl group alkylates and the ring opens while the P—N bond remains unbroken. Eventually, however, tepa is metabolized to phosphoric acid and a mixture of metabolites resulting from the amine portion of the molecule.

Most alkylating phosphorus chemosterilants contain aziridine rings without substituents on the ring carbons. Monomethyl substitution, however, often yields still active sterilants though their effectiveness is lower than that of their unsubstituted analogs (BOŘKOVEC 1969). Nevertheless, because toxicity of chemosterilants varies independently from sterilizing activity (Table II), a less active compound may be sometimes preferable to a more active one if the toxicological characteristics of the former are exceptionally favorable. Substituents other than the methyl group or a multiple substitution on the ring carbons usually lead to marginal sterilants or to entirely inactive compounds.

b) *Nonalkylating organophosphorus chemosterilants*

The replacement of aziridinyl groups in alkylating sterilants by alkylamino groups produces a new class of chemosterilants that, nevertheless, exhibit biological effects remarkably similar to those of alkylating agents (MORGAN 1967). Figure 5 shows two most important representatives of the group.

In distinction to the large number of active aziridines only a handful of the nonalkylating analogs are sterilants. The most effective substituent is a dimethylamino group perhaps because of its steric resemblance to the aziridinyl ring. Prior to the discovery of the sterilizing activity of hexamethylphosphoric triamide (hempa) all chemosterilants effective in male insects were alkylating agents. Also, since the physiological effects of hempa in the housefly were strikingly similar to those of the aziridine tepa (MORGAN and LABRECQUE 1964, MORGAN 1967), the suspicion that both compounds had a similar mode of action was difficult to resist. It was clear, however, that hempa and analogous compounds could not directly alkylate and react with a nucleophile. In fact, hempa was long being used as an excellent and relatively inert solvent in organic reactions and its hydrolytical

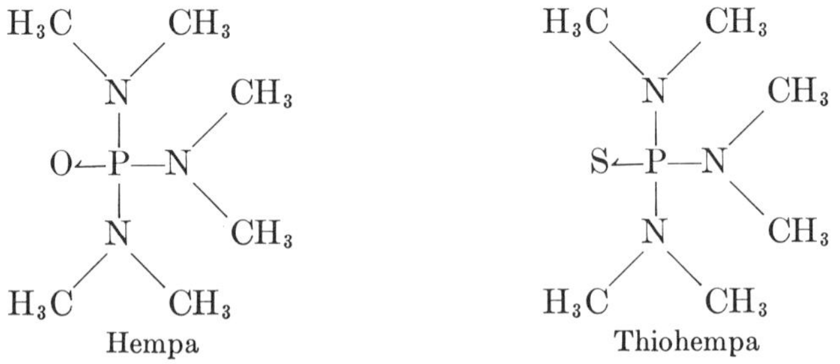

Fig. 5. Nonalkylating phosphorus chemosterilants.

stability in dilute acids and bases was well known. The first indications that hempa could conceivably act as an alkylating agent was given by the results of our study of its metabolism in insects (CHANG *et al.* 1967). The relatively simple metabolic pathway is shown in Figure 6.

The methylol compound, however, was never isolated or identified in the metabolic products, apparently because of its low stability.

Since, in the male housefly, the pentamethylphosphoric triamide was inactive as a sterilant, the activity must have been owing to its metabolic precursors. When finally the methylol intermediate was synthesized in a surprisingly simple reaction (TERRY and BORKOVEC 1970) and its activity was found equal to that of hempa, the implication of its direct sterilizing effects became stronger. However, exactly how a methylol amide would alkylate is still an open question. Formaldehyde, which the compound readily liberates, is a mild alkylating agent but formaldehyde itself has never shown any sterilizing activity when it was injected into the insect's hemolymph. It cannot be excluded, however, that the same agent liberated at the site of action may have a sterilizing effect and that hempa penetrates the cell membrane better than does formaldehyde. In analogy with the metabolism of certain dimethylcarbamates, schradan, and other insecticides, hempa is demethylated by the enzymes of housefly microsomes (AKOV *et al.* 1968). This mechanism implies the diffusion of hempa into a cell, and since its sterilizing effects are most probably genetic and involve the chromosomes, the metabolism and the alkylating reaction could conceivably take place within the same cell. In this respect an important and still unresolved question is whether the gametes, particularly the spermatozoa, contain microsomal enzymes capable of activating hempa. At the time of treatment, the testes of an adult male housefly contain mature spermatozoa that are transferred during

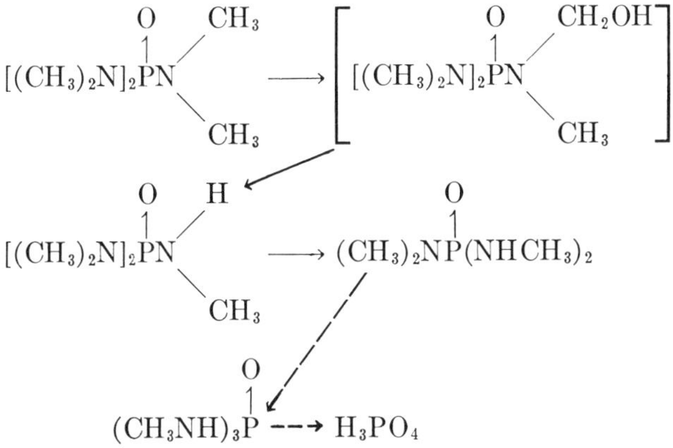

Fig. 6. Metabolism of hempa in the housefly.

copulation into the spermatheca of a female. Subsequently, the spermatozoon penetrates an egg in the female's oviduct and a zygote is formed while the egg is laid. This sequence of processes occurs whether the male was normal or treated with hempa. Only later, when embryogenesis proceeds to or beyond the blastoderm stage, the egg fertilized by the sterilized male stops developing whereas the egg fertilized by an untreated fertile male develops normally and gives rise to a larva. If the mechanism of sterilization were indeed intracellular, hempa would have to penetrate the sperm-cell wall shortly after treatment, be converted to the alkylating intermediate, and then react with the chromatin in the head portion of the spermatozoon. However, no experimental evidence supporting this mechanism is available at present.

That the ability of the entire organism to demethylate hempa is not directly related to the sterilant's activity follows from an experiment in which houseflies susceptible or resistant to Isolan® were injected with hempa and its metabolism and sterilizing activity were determined (AKOV and BOŘKOVEC 1968, CHANG and BOŘKOVEC 1969). Though hempa was metabolized twice as rapidly in the resistant flies than in the susceptible ones, its sterilizing activity was equal in both strains. This result appeared to be inconsistent with the hypothesis that hempa itself was inactive and that it required metabolic conversion to the active methylol intermediate. The original simple assumption was that in flies with a high metabolic capacity for demethylation, *i.e.*, those resistant to Isolan, hempa would be metabolized rapidly, the methylol intermediate's concentration would be high, and thus the sterilizing activity would be much higher than that in the susceptible flies. This proved to be an oversimplication perhaps because the concentration of the presumed active intermediate, *i.e.*, the methylol compound, could not be determined directly and the rate of its formation may not have been as important as its persistence.

Another attempt for gaining evidence for the metabolic activation hypothesis was based on the supposition that a synergist that is known to decrease the metabolic demethylation in insects should also decrease the sterilizing activity of hempa. However, when the synergist Tropital® was used in conjunction with hempa (CHANG and BOŘKOVEC 1969), the sterilizing activity of hempa remained unchanged.

In distinction to schradan and to other methylated amides whose methylol derivatives have cholinesterase inhibiting properties, the methylol analog of hempa and hempa itself are not known to affect this enzyme. Perhaps the low stability of the hempa derivative as compared, for example, to the relatively stable hydroxy derivative of schradan is responsible for the difference.

In rats, the metabolism of hempa is basically the same as in the housefly, and high doses of the compound produce sterility (JACKSON and CRAIG 1966). However, when hempa was administered continuously for 169 days to rats at a dose of two mg/kg/day, no apparent toxicological or reproduction effects were observed (SHOTT *et al.* 1971). In general, the toxicology of hempa in rats and rabbits (KIMBROUGH and GAINES 1966, SHOTT *et al.* 1971) does resemble that of tepa in qualitative or quantitative aspects. Bone marrow depression, a characteristic effect of tepa and of most other biological alkylating agents, was not observed with hempa. On the other hand, the high and often fatal incidence of lung diseases associated with the intake of hempa is not common after treatment with tepa. The relatively low acute oral toxicity of hempa to rats (LD_{50} 2,600 mg/kg) is also distinctly different from that of tepa (LD_{50} 37 mg/kg).

In the housefly the sterilizing activity of tepa is about 50 times as high as that of hempa but the toxicities of the two compounds are about equal (CHANG and BOŘKOVEC 1969). A small difference between sterilizing and toxic doses is an undesirable characteristic for a chemosterilant and from a practical standpoint tepa would have a clear advantage over hempa. However, since all insects are not equally susceptible to chemosterilants, generalizations concerning toxic and sterilizing doses cannot be made. On the other hand, the metabolism of hempa by a stepwise demethylation appears to be a general mechanism applicable to insects, vertebrates (JONES and JACKSON 1968), and even plants (HEATH *et al.* 1955). Unfortunately, quantitative relationship between the individual metabolites is known only in the housefly and whether the biological effects are associated with the persistence of any of the metabolites is not known. As mentioned earlier, pentamethylphosphoric triamide is the principal metabolite in the housefly and its demethylation is much slower than that of hempa. Also, its sterilizing activity is almost undetectable. Conceivably, a compound that would be demethylated faster than hempa could also be a more effective sterilant. To test this possibility, numerous analogs of hempa were synthesized (TERRY and BOŘKOVEC 1965, 1967, 1968, and 1970) and screened for sterilizing activity in houseflies and in other insects (KLASSEN *et al.* 1968, MADHUKAR *et al.* 1971, TERRY and CRYSTAL 1972, TERRY *et al.* 1972). However, since none of the compounds was more effective than hempa, metabolic studies were not conducted.

The outstanding characteristics of the PO, PS, or PN moieties are particularly evident in the nonalkylating phosphorus chemosterilants. Whereas the aziridinyl group when attached to a wide variety of polar systems, e.g., the carbonyl group, yields effective chemosterilants, only the PO, PS, and the CN groups (in *s*-triazines and certain dithiazoles) are suitable carriers for the dimethylamino substituent. Excellent solubility of most of the phosphorus chemosterilants in polar and nonpolar solvents may be an important factor but since many effective and poorly soluble chemosterilants are also known other characteristics of the phosphorus moiety may be contributing to the activity of the functional group. It is hoped that further synthesis, testing, and metabolic studies of phosphorus chemosterilants will increase our understanding of these characteristics.

Summary

Alkylating derivatives of phosphoric and phosphorothioic acids and cyclic azaphosphorines are chemical cytostatic and mutagenic agents with selective activity on gonadal tissues of insects. Structure-activity relationship studies revealed a primary role of the alkylating moiety and secondary role of the phosphorus carrier moiety. Toxicological properties of this class of chemosterilants are basically different from those of phosphorus insecticides. Nonalkylating organophosphorus chemosterilants show similarities in mode of action that implicate possible activation step as a condition for biological activity. Structure-activity correlations indicate species-specific differences in metabolic patterns and much narrower range of activity as compared with the alkylating analogs. Experiments with synergists and with resistant strains of insects revealed some basic differences between the biological activity of phosphorus chemosterilants and insecticides.

References

AKOV, S., and A. B. BOŘKOVEC: Metabolism of the chemosterilant hempa by carbamate-resistant house flies. Life Sci. **7**,(II), 1215 (1968).

——, J. E. OLIVER, and A. B. BOŘKOVEC: *N*-Demethylation of the chemosterilant hempa by house fly microsomes. Life Sci. **7**(II), 1207 (1968).

BEROZA, M., and A. B. BOŘKOVEC: The stability of tepa and other aziridine chemosterilants. J. Med. Chem. **7**, 44 (1964).

BOŘKOVEC, A. B.: Sexual sterilization of insects by chemicals. Science **137**, 1034 (1962).

—— Insect chemosterilants. New York: Interscience (1966).

—— Alkylating agents as insect chemosterilants. Ann. New York Acad. Sci. **163**(Art. 2), 860 (1969).

——, R. L. FYE, and G. C. LaBrecque: Aziridinyl chemosterilants for houseflies. *U.S. Department of Agriculture,* Agricultural Research Service **33-129**, 1 (1968).

——, C. W. WOODS, and R. T. BROWN: Insect chemosterilants. III. 1-Aziridinylphosphine oxides. J. Med. Chem. **9**, 522 (1966).

CHANG, S. C., and A. B. BOŘKOVEC: Structure-activity relationship in analogs of tepa and hempa. J. Econ. Entomol. **59**, 1359 (1966).

—— —— Comparative metabolism of ^{14}C-labeled hempa by house flies susceptible and resistant to Isolan®. J. Econ. Entomol. **62**, 1417 (1969).

——, P. H. TERRY, C. W. WOODS, and A. B. BOŘKOVEC: Metabolism of hempa uniformly labeled with C^{14} in male house flies. J. Econ. Entomol. **60**, 1623 (1967).

——, C. W. WOODS, and A. B. BOŘKOVEC: Sterilizing activity of bis(1-aziridinyl)phosphine oxides and sulfides in male house flies. J. Econ. Entomol. **63**, 1744 (1970).

HEATH, D. F., D. W. J. LANE, and P. O. PARK: The decomposition of some organophosphorus insecticides and related compounds in plants. Trans. Roy. Soc. (London) **239B**, 191 (1955).

JACKSON, J., and A. W. CRAIG: Antifertility action and metabolism of hexamethylphosphoramide. Nature **212**, 86 (1966).

JONES, A. R., and H. JACKSON: The metabolism of hexamethylphosphoramide and related compounds. Biochem. Pharmacol. **17**, 2247 (1968).

KIMBROUGH, R., and T. B. GAINES: Toxicity of hexamethylphosphoramide in rats. Nature **211**, 146 (1966).

KLASSEN, W., J. F. NORLAND, and A. B. BOŘKOVEC: Potential chemosterilants for boll weevils. J. Econ. Entomol. **61**, 401 (1968).

KNIPLING, E. F.: Sterile-male method of population control. Science **130**, 902 (1959).

—— Some basic principles in insect population suppression. Bull. Entomol. Soc. Amer. **12**, 7 (1966).

MADHUKAR, B. V. R., M. K. K. PILLAI, and A. B. BOŘKOVEC: Chemosterilization of the yellow-fever mosquito. 2. Laboratory evaluation of phosphoramides by larval and pupal treatments. J. Econ. Entomol. **64**, 1027 (1971).

MORGAN, P. B.: Effects of hempa on the ovarian development of the house fly, *Musca domestica* (Diptera: Muscidae). Ann. Entomol. Soc. Amer. **60**, 812 (1967).

——, and G. C. LaBrecque: Effect of tepa and metepa on ovarian development of house flies. J. Econ. Entomol. **57**, 896 (1964).

SHOTT, L. D., A. B. BOŘKOVEC, and W. A. KNAPP, JR.: Toxicology of hexamethylphosphoric triamide in rats and rabbits. Toxicol. Applied Pharmacol. **18**, 499 (1971).

TABER III, S., and A. B. BOŘKOVEC: Chemical sterilization of honey bee spermatozoa *in vitro*. Nature **224**, 1217 (1969).

TERRY, P. H., and A. B. BORKOVEC: Phosphoramides and thiophosphoramides as insect chemosterilants. U.S. Pat. 3,205,130 (1965).

—— —— Insect chemosterilant. IV. Phosphoramides. J. Med. Chem. **10**, 118 (1967).

—— —— Insect chemosterilants. VI. Oxidation of hexamethylphosphoric triamide and the synthesis of *N*-formylphosphoramides. J. Med. Chem. **11**, 958 (1968).

—— —— Insect chemosterilants. IX. *N*-(Hydroxymethyl)-*N,N′,N′,N″,N″* pentamethylphosphoric triamide. J. Med. Chem. **13**, 782 (1970).

—— ——, and D. G. MCHAFFEY: Chemosterilants against the boll weevil. 3. Phosphorus amides. J. Econ. Entomol. **65**, 1550 (1972).

——, and M. M. CRYSTAL: Chemosterilants against screwworm flies. 2. J. Econ. Entomol. **65**, 307 (1972).

Manuscript received September 4, 1973; accepted October 17, 1973.

New aspects of organophosphorus pesticides. VII. Metabolism, biochemical, and biological aspects of Nemacur® and related phosphoramidate compounds.

By

T. BILL WAGGONER* and A. M. KHASAWINAH*

Contents

* Chemagro Division of Baychem Corporation, Research and Development, P.O. Box 4913, Kansas City, Missouri 64120.

I. Introduction

The purpose of this review is to discuss the biological activity of some new organophosphorus nematicides. Structure-activity relationships are given for several *N*-substituted phosphoramidates, and further biological studies are presented for the most active compound, typical of this series of phosphoramidates and its major metabolites. The chemistry and metabolism of the biologically active compounds are discussed in relation to their nematicidal activity.

a) N-Substituted phosphoramidates

Discussion is limited to a series of phosphoramidates where only the nitrogen substituents are different. Examples are presented for *N*-methyl and *N*-isopropyl phosphoramidates.

1. *N-Methyl analogs.* — The structures of these compounds are in Table I. Some data are available for their nematicidal activity from initial screening tests, and, therefore, the effect of mono- and dimethyl substituted amidic acid on their biological activity is discussed. In general, this series is less active than the corresponding *N*-isopropyl analogs.

2. *N*-Isopropyl analogs. — A single isopropyl substituent on the amidic acid results in excellent nematicidal activity. Considerably more information has been obtained for this compound since it has been developed further as a nematicide under the name Nemacur®[1] [ethyl 3-methyl-4-(methylthio)phenyl (1-methylethyl)phospora-

Table I. *Oral* LD_{50} *toxicity of N-methyl and N-isopropyl phosphoramidate compounds to rats* (HOFFMANN 1972).

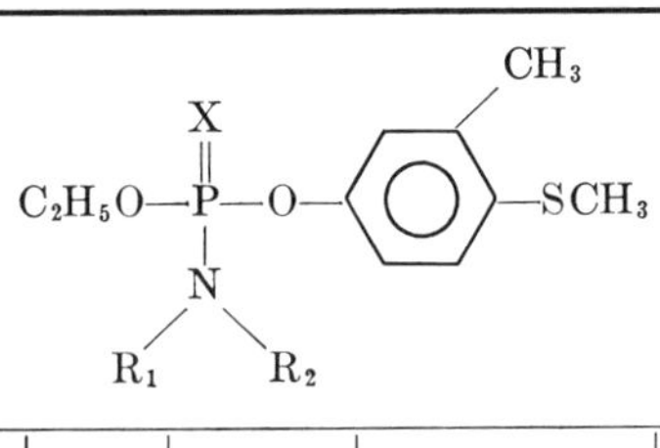

Compound no.	X	R_1	R_2	LD_{50} (mg/kg)
SRA 3511	O	H	CH_3	2–5
SRA 3147	O	H	H	5–10
SRA 3512	O	CH_3	CH_3	10–25
SRA 3886[a]	O	H	$(CH_3)_2CH$	10–25
SRA 2090	S	H	$(CH_3)_2CH$	25–50

[a] Nemacur®, ethyl 3-methyl-4-(methylthio)phenyl (1-methylethyl)phosphoramidate.

Table II. *Inhibition of acetylcholinesterase by Nemacur® and its metabolites found in plants* (WAGGONER 1972).

Compound	I_{50} (molar conc.)	I_{50} (pg/mg enzyme[a])	Rel. inhibition (% of Nemacur®)
Nemacur®	18.0×10^{-9}	200	100
Nemacur sulfoxide	9.7×10^{-9}	120	167
Nemacur sulfone	9.7×10^{-9}	120	167
Unknown[b]	3.6×10^{-9}	30	670

[a] Horse serum acetylcholinesterase, Type IV.
[b] An intact phosphoramidate which occurs briefly in plants as a minor metabolite.

midate]. Nematicidal activity has been determined for the corresponding PS analog (SRA 2090) and thiooxidation products, Nemacur sulfoxide and Nemacur sulfone. The thiooxidation products have subsequently been found as the major metabolites of Nemacur.®.

b) *Toxicology*

From subacute toxicological studies, a "no-effect" level of four ppm to rats and one ppm to dogs has been established for Nemacur®. The *N*-substituted phosphoramidates described in this paper are classified as highly toxic and are potent inhibitors of acetylcholinesterase.

1. Acute oral LD_{50}. — Toxicity to rats is summarized in Table I. Addition of one methyl (SRA 3511) lowers the LD_{50} by 50 percent compared to the amidic acid (SRA 3147). Several factors may cause this. One explanation is that the basicity of the nitrogen is increased due to electronic effects of the methyl. Using the same reasoning for addition of a second methyl (SRA 3512), the LD_{50} expected would still be lower. However, a dimethyl substituent results in a tertiary nitrogen and steric considerations may now be the limiting factor. Obviously, something has occurred to give an LD_{50} value approximately twice as high as the unsubstituted amidic acid. The effect of one isopropyl group (SRA 3886) results in the same LD_{50} value as the dimethyl compound. Steric factors may also account for this and, therefore, the overall toxic effect for both compounds is the same. The PS analog (SRA 2090) is less toxic than the PO analog as is normally observed for PS and PO compounds.

2. Inhibition of acetylcholinesterase. — The effect on blood cholinesterase for Nemacur® and intact phosphoramidate metabolites is shown in Table II. All metabolites are significantly more active

[1] Nemacur is registered trademark of the Parent Company of Farbenfabriken Bayer GmbH, Leverkusen.

than the parent compound. The unknown metabolite isolated from plants has been previously described and consists of an intact phosphoramidate. It has beeen found within 14 days after treatment but disappears rapidly and does not contribute to terminal toxic residues.

II. Metabolism and biochemical properties

Chemical properties and degradation in biological systems have been studied for Nemacur® and are discussed as they relate to efficacy. Behavior in animal systems is also given to illustrate its rapid breakdown compared to relative persistence in soil and plants.

a) Water

Nemacur® is fairly stable at pH 5 to 7 at 30° to 50°C and does not hydrolyze over a 20-day period when dissolved in phosphate buffers protected from light. However, hydrolysis is observed in natural pond water at pH 7 and 29°C under field conditions in the presence of silt. Approximately 50 percent is hydrolyzed within five days under these natural outdoor conditions.

Under basic conditions the rate of hydrolysis increases. Approximately 50 percent hydrolyzes within 13 days in a phosphate buffer of pH 9 at 30°C protected from light. Rates of hydrolysis for Nemacur®, Nemacur sulfoxide, and Nemacur sulfone are summarized in Table III. Thiooxidation products hydrolyze three to four times faster than Nemacur® and pK values for corresponding phenols are in Figure 1. The pK of 10.5 for Nemacur® is similar to phenol of pK 10.0 at 25°C (STECHER 1960). Thiooxidation produces more acidic phenols which correlate with the faster rates of hydrolysis of the corresponding phosphoramidates.

Investigation of acid-catalyzed hydrolysis of a series of 2,4-dichlorophenyl methyl *N*-alkylphosphoramidates has shown that cleavage occurs mainly at the P—N bond, and 2,4-dichlorophenol is a minor hydrolysis product (GARRISON and BOOZER 1968). Under basic

Table III. *Hydrolysis of Nemacur® and metabolites* (HOFER 1972).[a]

Compound	$t\frac{1}{2}$ (min.)	Rate constant K (min.$^{-1}$)
Nemacur®	870	0.79×10^{-3}
Nemacur sulfoxide	300	2.3×10^{-3}
Nemacur sulfone	215	3.2×10^{-3}

[a] Solvent = isopropyl alcohol:water (1:1), pH 11.4, 37°C.

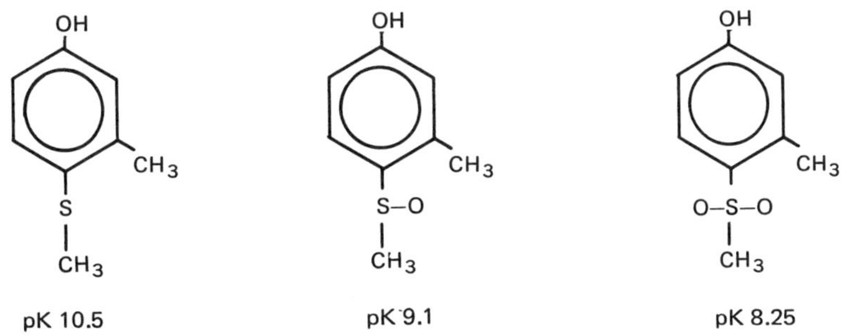

Fig. 1. Phenols from Nemacur® and its thiooxidation products (HOFFER 1972).

conditions, cleavage is mostly at the P—O bond and the major products from Nemacur® and thiooxidized analogs are the phenols.

b) Soil

The major route of degradation of Nemacur® is thiooxidation producing Nemacur sulfoxide as the major metabolite in soil (see Fig. 2). Nemacur® is oxidized rapidly and disappears completely in about three weeks which corresponds to the time of maximum concentration of the sulfoxide. Nemacur sulfone accumulates slowly over a 12-week period but its amount relative to the sulfoxide is much less.

c) Plants

Metabolism of Nemacur® in tomato, bean, and peanut plants has been previously reported (WAGGONER 1972). The route of administration was stem-injection and further studies are now discussed for soil treatments.

All plant studies have been conducted with Nemacur® labeled with carbon-14 and tritium so that metabolites, consisting of intact phosphoramidate structures, are readily identified from their tritium/carbon-14 ratios.

1. Extraction procedure. — For isolation of Nemacur® and metabolites, initial extraction with acetone and subsequent extraction and partitioning with chloroform resulted in an organosoluble fraction containing mostly unconjugated metabolites and a water fraction containing conjugates (see Fig. 3). The water fraction was further investigated using gel permeation and thin-layer chromatography (TLC) and hydrolysis procedures. The activity remaining in the plant solids was further investigated by hydrolysis and TLC techniques. Identification of metabolites was confirmed by infrared and mass spectral data with known standards.

2. Major metabolites. — Nemacur sulfoxide and Nemacur sulfone have been identified as the principal significant metabolites in plants

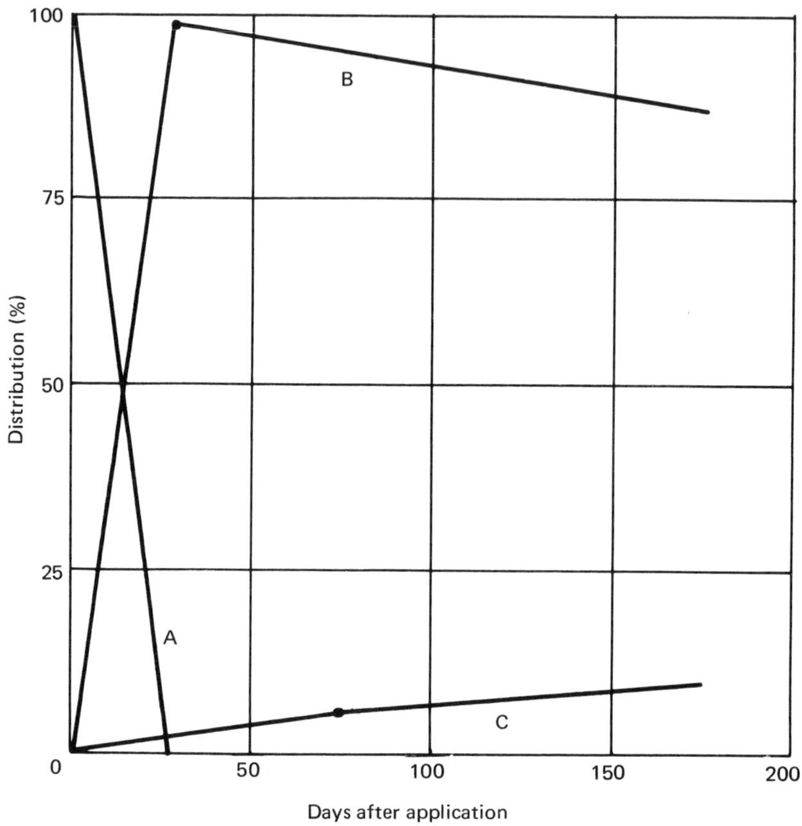

Fig. 2. Distribution of Nemacur® (*A*), Nemacur sulfoxide (*B*), and Nemacur sulfone (*C*) in soil.

(WAGGONER 1972). For all studies, total radioactive residue consisted not only of significant intact phosphoramidate residues, but also non-significant hydrolysis fragments. The results of these studies for total residue are briefly discussed for the three fractions resulting from the extraction procedure.

α) *Organosoluble.* Intact phosphoramidates found were Nemacur®, Nemacur sulfoxide, Nemacur sulfone, and two minor metabolites which occurred in the first two weeks after treatment, each contributing to approximately one-to-two percent of the total residue in the plant. The cholinesterase effect of one of them was shown in Table II, but no further information is known about the other one. The free thiooxidized phenols were found but none of the parent phenol.

A sixth component was the least polar, containing only carbon-14, and was associated with green pigments and could be further sepa-

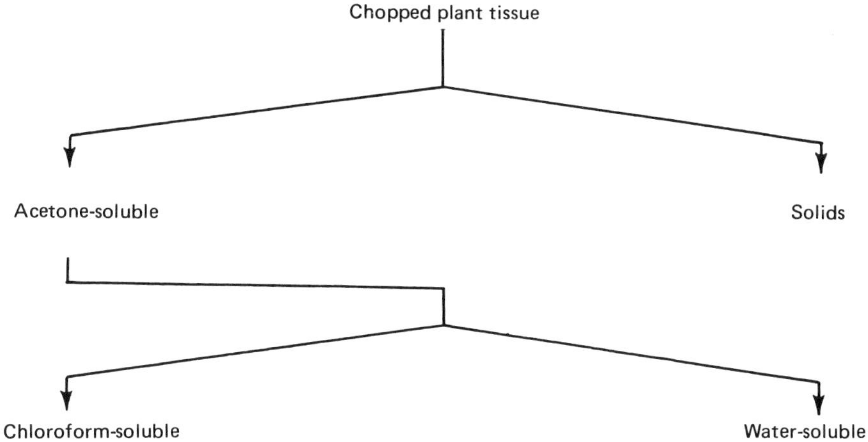

Fig. 3. Extraction scheme for Nemacur® and metabolites from plant tissue.

rated into three fractions on Sephadex LH-20, all having a molecular weight greater than 1,000. These relatively nonpolar compounds were found in all plants studied. The remaining seventh component in the chloroform fraction was the most polar and resembled the metabolites in the water fraction.

β) *Water-soluble.* Hydrolysis with either β-glucosidase or acid converted nearly all the tritium activity to organosoluble products. This accounted for all tritium activity in the water fraction which contained zero-to-negligible amounts of carbon-14 activity. The water-soluble conjugates were separated and isolated on Sephadex G-10 by column chromatography and enzymatically hydrolyzed to the phenols of the corresponding sulfoxide and sulfone analogs of Nemacur®.

γ) *Extracted plant solids.* Acid hydrolysis produces organo-soluble products which have been identified as phenols identical to those found in the water fraction. The remainder of the radioactivity in the solids was mainly due to carbon-14. Tritium/carbon-14 ratios lower than the ratio for Nemacur® indicated degradation of the phosphoramidate structure and, therefore, residues in the plant solids were considered as nonsignificant.

3. Application by stem-injection. — Recovery of radioactivity administered in previous studies (WAGGONER 1972) was approximately ten percent 28 days after treatment. The fate of the unaccountable radioactivity has been investigated further.

α) *Recovery of radioactivity.* Bean plants were grown in a glass chamber equipped for trapping volatile products which may form. The plants were stem-injected with Nemacur® labeled in the ethyl

Table IV. *Recovery of total carbon-14 and tritium radioactivity from bean plants.*

Fraction	Ethyl-^{14}C 3-methyl-4-(methyl-^{3}H-thio)phenyl (1-methylethyl)phosphoramidate[a] — Percent of administered dose		
	^{14}C	^{3}H	^{3}H/^{14}C
Chloroform-soluble	11.0	32.2	26.7
Water-soluble	2.1	15.2	61.0
Extracted plant solids	29.1	23.8	7.4
Volatiles in acid trap	0.0	7.4	—
Volatiles in base trap	38.2	0.0	—
Total	80.4	79.6	—
	Ethyl 3-methyl-4-(methyl-^{3}H-thio) phenyl-^{14}C(1-methylethyl)phosphoramidate[b]		
Chloroform-soluble	35.4	35.4	4.1
Water-soluble	42.0	42.0	4.1
Extracted solids	24.8	24.8	4.1
Total	102.2	102.2	—

[a] ^{3}H/^{14}C = 9.0.
[b] ^{3}H/^{14}C = 4.1.

group with carbon-14 and with tritium in the methylthio substituent of the aromatic ring. The plants were maintained under controlled conditions for four weeks. The results are summarized in Table IV.

Total recovery of ethyl-^{14}C/^{3}H activity was approximately 80 percent which compared to ten percent recovery from field tests. Nearly half the carbon-14 activity was found in the NaOH trap as carbon dioxide from the ethyl ester portion of the phosphoramidate. The acid trap contained only tritium, probably in free phenols which were volatile. Water was ruled out because distillation of the aqueous fraction failed to produce tritiated water.

The test was repeated with Nemacur® labeled in a different position with carbon-14 in the phenyl ring and tritium in the same methylthio substituent. Recovery was quantitative as shown in Table IV and the phenol ring remained intact in all fractions as shown by a constant carbon-14 tritium ratio. This is also confirming evidence that tritium exchange did not occur, nor was the methylthio substituent cleaved from the ring.

β) *Distribution of metabolites.* The primary toxic metabolites were Nemacur sulfoxide and Nemacur sulfone and their relative distribution was determined for stem-injected plants (see Fig. 4). Sulfoxide built up rapidly with a corresponding drop in Nemacur® concentration. Nemacur sulfone reached a maximum after about a week and disappeared after about four weeks. Terminal residues consisted mainly of Nemacur sulfoxide, and foliar treatments in the field will most likely give this distribution.

4. Application by soil treatment. — For treatment of field and vegetable crops, Nemacur® is usually applied directly to the soil before planting. Residues in plants will then occur from absorption through the roots. Metabolites occur from conversion of Nemacur® in both the soil and the plants.

α) *Distribution of metabolites.* Plants growing in soil treated with Nemacur showed a different distribution of significant metabolites than observed from stem-injection. An example of a typical pattern is represented for tobacco in Figure 5. Initial residues of Nemacur® dropped rapidly with a corresponding increase of sulfoxide and slower build-up of sulfone. This contrasts with the pattern for stem-injected plants where sulfone reached a maximum within a week and then declined.

β) *Degradation in root crops.* A uniform distribution of total radioactive residue is shown in Figure 6 for carrots as a typical root crop growing in treated soil. Carrots showed a fairly uniform distribution of carbon-14 and tritium. However, Nemacur® was degraded and the residue in the roots consisted of conjugated thiooxidized phenols and nonsignificant radioactive residues which result from incorporation into naturally-occurring plant substances. Under normal field conditions, the combined residue of Nemacur® and its sulfoxide and sulfone in carrots was less than 0.1 ppm.

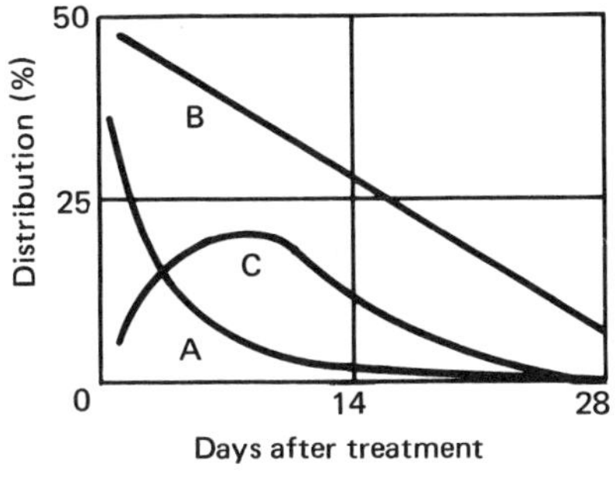

Fig. 4. Distribution of Nemacur® (*A*), Nemacur sulfoxide (*B*), and Nemacur sulfone (*C*) in plants after treatment by stem-injection.

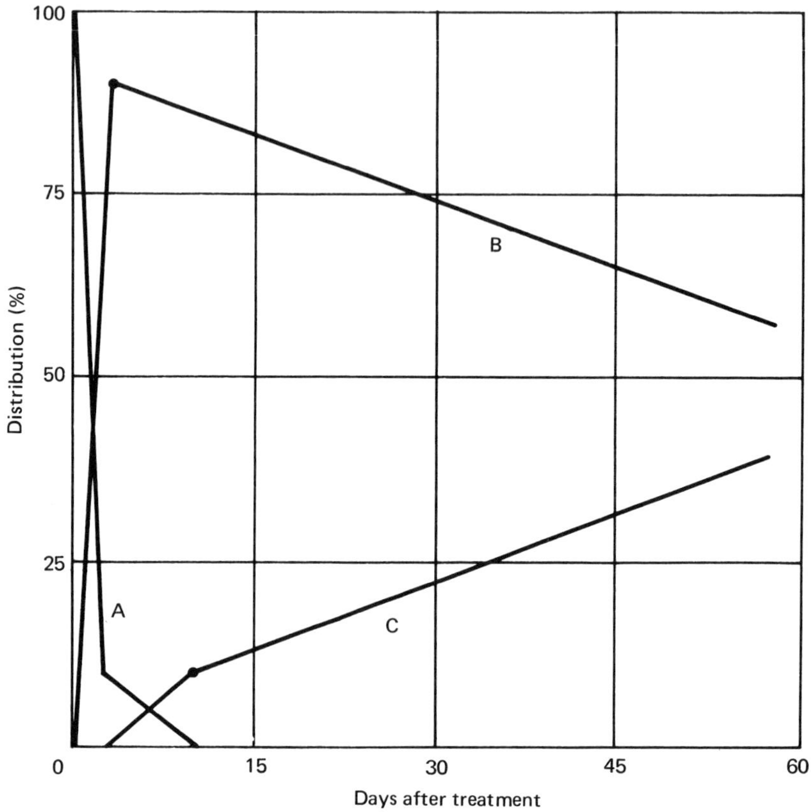

Fig. 5. Distribution of Nemacur® (*A*), Nemacur sulfoxide (*B*), and Nemacur sulfone (*C*) in plants after treatment of soil.

d) *Animals*

Major routes of metabolism in animals consisted of hydrolysis and thiooxidation, the same as found for plants and soil. Rapid degradation and excretion from either single or multiple doses did not result in accumulation of residues in tissues of rats and meat and milk of dairy cattle.

1. Rats. — Single oral doses of Nemacur® labeled with carbon-14 at two positions and tritium at one position were given to rats. Recovery of all radioactivity was attempted and the results summarized in Table V.

α) *Excretion.* Approximately 40 to 50 percent of the administered activity was eliminated as carbon dioxide which indicates a rapid hydrolysis of C—O—P and C—N—P esters. The remainder of carbon-14 activity and most of the tritium activity was excreted in the

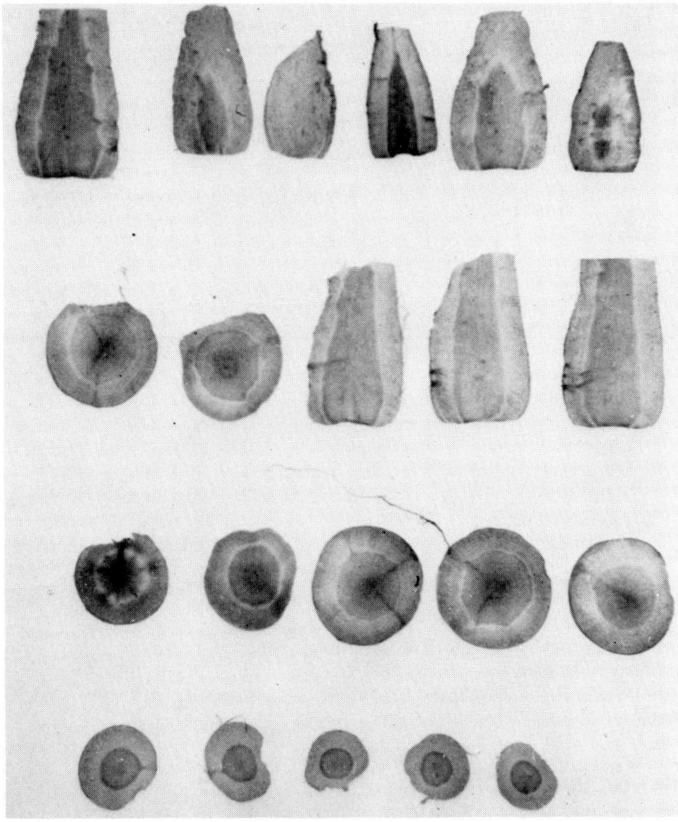

Fig. 6. Autoradiogram showing the uniform distribution of radioactivity in a mature carrot after treatment of soil with Nemacur® at planting.

urine with smaller amounts in the feces. Most of the radioactivity was eliminated in the urine or expired air within 12 to 15 hours after treatment.

β) *Major metabolites.* Within the 48-hour period after treatment, the only intact phosphoramidate present was a trace amount of Nemacur sulfoxide in the urine. Nearly all the residue in urine consisted of polar metabolites of a molecular weight range of 400 to 800 mwu.

Nemacur® and its sulfoxide and sulfone analogs occurred in liver immediately after treatment but degraded within 48 hours. *In vitro* studies with a microsomal fraction from liver showed a very rapid degradation of Nemacur® in ten minutes. An *N*-dealkylated phosphoramidate, identified by comparison of mass spectral data with a known standard, occurred as a minor metabolite in the microsomal fraction but disappeared within 15 minutes.

Table V. *Excretion and residues of carbon-14 and tritium from double-labeled Nemacur®.*[a]

Route	Percent of administered dose			
	Ethyl-^{14}C (methyl-^{3}H-thio)		1-Methylethyl-^{14}C (methyl-^{3}H-thio)	
	^{14}C	^{3}H	^{14}C	^{3}H
Tissue and excretion				
Urine	40.4	95.5	12.2	79.3
Feces	6.0	6.8	4.5	11.8
Blood	0.1	0.1	0.1	0.1
Liver	7.3	0.1	3.5	0.1
Total body	6.0	2.2	7.6	2.2
Traps for volatiles				
Ethanolamine	39.7	0.0	49.7	0.0
Acid	0.8	0.0	17.0	0.0
Total recovery	100.4	104.6	93.6	93.4

[a] Ethyl-^{14}C 3-methyl-4-(methyl-^{3}H-thio)phenyl (1-methylethyl)phosphoramidate and ethyl 3-methyl-4-(methyl-^{3}H-thio)phenyl (1-methylethyl-^{14}C)phosphoramidate; rats were given a single oral dose of two mg/kg and sacrificed 48 hours after treatment.

2. Dairy cattle. — Degradation of Nemacur® was further demonstrated in animals by the lack of accumulation in meat tissues and milk. Using a gas chromatographic analytical residue method (THORNTON 1972), it was shown that Nemacur® and sulfoxide and sulfone metabolites did not accumulate after animals had been given daily doses of Nemacur® for 30 days at 1.2 mg/kg body weight which is equivalent to approximately 40 ppm in the total diet. Residues were less than 0.01 and 0.001 ppm for tissues and milk, respectively.

III. Efficacy against soil parasitic nematodes

Nematicidal activity has been determined for *N*-methyl analogs from laboratory studies. Nemacur metabolites and the PS analog have been studied further in laboratory and field studies. For initial studies parasitic nematodes and proper testing procedures were selected to give as realistic test as possible. A detailed explanation of laboratory test procedures has been described (HOMEYER 1971). A brief discussion will be given here.

a) Target organisms

Several hundred species of nematodes are known to attack plants. Numerous species are controlled by Nemacur®, but for testing procedures a few species are chosen to represent the many types that are encountered in the field.

1. Endoparasites. — Some nematode species enter the root completely or are attached to it and lay their eggs and reach maturity within the root. Examples used in these tests consist of the root-knot nematode, *Meloidogyne* spp. and the lesion nematode, *Pratylenchus* spp.

2. Ectoparasites. — Other types of nematodes are free-living and actually feed without entering the root. These nematodes are also included in the field tests reported here.

b) Test procedures

All tests were conducted in soil the natural habitat of nematodes. Percentage of effectiveness for endoparasites was determined against untreated checks using known methods (Abbott 1925). The effect on freeliving nematodes was determined by known techniques utilizing flotation and microscopic examination.

1. Laboratory. — The test procedure consisted of mixing the active ingredient with soil infested with the test organism and placing this mixture in half-liter pots in the laboratory. The compounds were allowed to act for one week at 20°C. Efficacy was measured against untreated controls. This procedure was followed to obtain the data presented in Tables VI through VIII.

2. Field. — The procedure for Table IX and X consisted of four replicates. Test crops were planted one day after treatment. Soil was sampled twice for analysis of free-living nematodes. Plants were evalu-

Table VI. *Structure-activity relationships of N-methyl and N-isopropyl phosphoramidate compounds* (Homeyer 1972).

Compound	*N*-substituent	Conc. in soil for 100 % control[a] (ppm)
SRA 3147	$—NH_2$	50
SRA 3511	$—NHCH_3$	>50
SRA 3512	$—N(CH_3)_2$	40
SRA 2090[b]	$—NHCH(CH_3)_2$	10
Nemacur®	$—NHCH(CH_3)_2$	5

[a] Laboratory test, *Meloidogyne* spp.
[b] PS analog of Nemacur®.

Table VII. *Activity of Nemacur®, PS analog, and metabolites as effect of concentration in soil on percent control* (HOMEYER 1972).[a]

Compound	Control (%) Conc. in soil (ppm) 5	2.5	1.25
Nemacur PS analog	95	90	30
Nemacur®	100	98	95
Nemacur sulfoxide	100	95	60
Nemacur sulfone	98	75	0

[a] Laboratory test, incubated one week, 20°C.

Table VIII. *Duration of effectiveness of Nemacur®, its PS analog, and metabolites in soil* (HOMEYER 1972).[a]

Compound	Effectiveness in weeks at 10 ppm	5 ppm	2.5 ppm
Nemacur PS analog	12	8	5
Nemacur®	20	15	10
Nemacur sulfoxide	18	15	10
Nemacur sulfone	10	6	4

[a] Laboratory test with *Meloidogyne incognita.*

ated, based on yield, final plant weight, and root damage to determine root-knot damage which has been previously described (ZECK 1971).

c) *Effect on nematode populations*

The structure-activity relationships for *N*-methyl and *N*-isopropyl analogs are shown in Table VI. The effect of methyl substitution was relatively small as shown by the concentration given in ppm needed to give 100 percent control. The effect of isopropyl substitution was more pronounced. Nemacur was clearly the superior nematicide compared to unsubstituted and methyl analogs. The PS analog (SRA 2090) of Nemacur® was clearly less effective than Nemacur.® The

Table IX. *Effectiveness of Nemacur® and metabolites in a field test* (ZECK 1972).[a]

Type of nematode		Effectiveness (%)		
		Nemacur®	Nemacur sulfoxide	Nemacur sulfone
Free-living				
Weeks after treatment	6	100	100	79
	14	87	55	40
Root-knot				
Test crops[b] Okra		80	22	52
Squash		90	21	73

[a] Rate of treatment = 25 kg/ha, broadcast and incorporated 15 cm.
[b] Effectiveness determined at harvest (14 weeks).

Table X. *Effect of Nemacur® and metabolites on populations of nematodes in soil* (ZECK 1972).[a]

Compound	No.[b] of genera/200 g of soil			
	Tylenchus (ectoparasite)		*Meloidogyne* (endoparasite)	
	6	14	6	14
Check	50	192	116	1,751
Nemacur®	0	8	0	210
Nemacur sulfoxide	0	49	0	681
Nemacur sulfone	30	71	6	918

[a] Rate of treatment = 25 kg/ha, broadcast and incorporated 15 cm.
[b] Number of nematodes determined six and 14 weeks after treatment.

effect of Nemacur® was further investigated and efficacy for its major metabolites is also presented since they represent most of the residue found in soil and plants.

The PS analog of Nemacur® is included for comparison only, although its metabolic picture has not been studied. One would assume that Nemacur PS analog will be oxidized to PO in soil and plants, and that biological activity will be similar. However, the results of biological testing presented here show that the PO analog, Nemacur®, is the superior nematicide.

It is shown in Table VI that the concentration needed in soil to give 100 percent control was five ppm for Nemacur®. In Table VII, results are shown for lower concentrations in soil. It should be emphasized that Nemacur® is metabolized in soil mainly to its sulfoxide and smaller amounts of sulfone.

Nemacur® maintained relatively good control at the lowest concentration studied. During the period of one week for this test, approximately half the Nemacur® had most likely been converted to the sulfoxide as shown by separate soil metabolism studies and conversion to the sulfone would have been slight. Apparently, the PS analog was metabolized in soil by a different route than Nemacur® as suggested by the results in Table VII showing its greater loss of activity compared to Nemacur® and Nemacur sulfoxide.

The duration of effectiveness in soil is summarized in Table VIII and is the time required to decrease from 100 to zero percent control. The persistence of the PS analog and Nemacur sulfone was considerably less than Nemacur® and its sulfoxide. The similarity between the latter two compounds is readily seen and understandable since the major metabolic route of Nemacur was to the sulfoxide.

Effectiveness under field conditions is summarized in Table IX as measured six and 14 weeks after treatment. Persistence of Nemacur® was greater than either of its metabolites and was consistent with the results of the laboratory tests given in Table VIII. Nemacur sulfoxide was more active than the sulfone against freeliving (ectoparasites) nematodes, but was considerably less active against root knot (endoparasites) nematodes at harvest. Activity of sulfone against endoparasites can result from a slower degradation rate in plants, as previously shown (see Fig. 5).

The actual population count for these studies is summarized in Table X. Nemacur® and its sulfoxide showed 100 percent control at the shorter interval after treatment for both endo- and ectoparasitic nematodes compared to an untreated check. At the longer interval, Nemacur® was 88 and 95 percent effective, respectively, for the two types of nematodes. Both metabolites were less effective than Nemacur® at the longer interval.

d) Dispersibility of Nemacur® and analogs in soil

An important factor for the control of root-parasitic nematodes is the movement and persistence of the compound in soil. Water-soluble nematicides disperse more slowly than fumigants, for example, and must persist longer in soil. The dispersibility of an active ingredient is the depth (given in cm) at which the compound gives 95 percent kill of the nematode population within the test period. The dispersibility of the compounds of interest is given in Table XI.

The test consisted of applying the active ingredient to the top one

Table XI. *Dispersibility as depth of penetration of Nemacur®, PS analog, and metabolites in soil* (HOMEYER 1972).[a]

Compound	Soil penetration (cm)[b]
Nemacur PS analog	5
Nemacur®	8
Nemacur sulfoxide	12
Nemacur sulfone	10

[a] Rate of treatment is 10 g active ingredient per square meter.
[b] Depth determined by bioassay with *Meloidogyne incognita.*

cm of soil at two g/m^2 in special pots (Mitscherlich), and 12 mm of water were added each time to give an equivalent amount of rainfall of 1,200 mm/year. After four weeks two-cm thick layers of soil were removed and put into another pot and sown in lettuce. After an additional four weeks, the roots were assessed for galling.

The penetration was greatest for sulfoxide, less for sulfone, and least for Nemacur®. The penetration of the PS analog was the least of all. The depth of penetration was related to the polar characteristics and correlated with the position of the compounds on thin-layer silica gel plates where Nemacur® moved the farthest (least polar), followed by the sulfone and then the sulfoxide (most polar).

IV. General metabolic pathway

A general metabolic pathway for Nemacur® in soil, plants, and animals is presented in Figure 7.

Summary

A general metabolic pathway for Nemacur in soil, plants, and animals is described. Major degradation routes are thiooxidation, dealkylation, hydrolysis, and subsequent conjugation of phenols. Significant residues in soil and plants consist of Nemacur® and Nemacur sulfoxide in the early stage of degradation, but after the disappearance of the parent compound, the significant terminal residue consists of mainly Nemacur sulfoxide and lesser amounts of Nemacur sulfone. The remainder of the residue consists of conjugated phenols and fragments which have been incorporated into naturally-occurring plant materials. Degradation in animals is rapid and nearly all the residue is excreted in the first ten to 15 hours. Residues do not accumulate in tissues or milk of cattle.

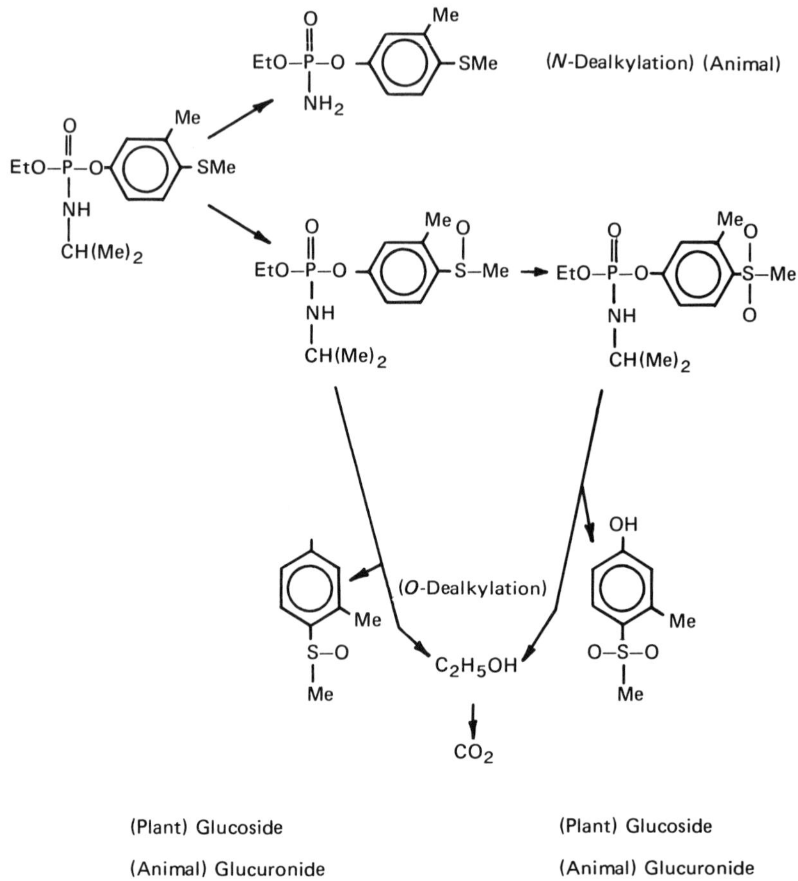

Fig. 7. General metabolic pathway of Nemacur® in soil, plants, and animals.

Nemacur® has been shown as a more active nematicide against parasitic nematodes than the corresponding *N*-methyl substituted phosphoramidate analogs. Biological activity and the duration of effectiveness in soil for the major metabolite, Nemacur sulfoxide, is similar to Nemacur®. The effect of Nemacur®, Nemacur sulfoxide, and Nemacur sulfone on soil populations of ectoparasitic nematodes is related to the persistence of the compound in soil; that is, Nemacur® is the most effective and has the highest persistence, whereas Nemacur sulfone is the least effective and is the least persistent. Against endoparasitic nematodes Nemacur sulfone is more effective than Nemacur sulfoxide, and this may be a result of the higher concentration of Nemacur sulfone in the roots.

References

ABBOTT, W. S.: A method of computing the effectiveness of an insecticide. J. Econ. Entomol. **18**, 265 (1925).

GARRISON, A. W., and C. E. BOOZER: The acid-catalyzed hydrolysis of a series of phosphoramidates. J. Amer. Chem. Soc. **90**, 3486 (1968).

HOFER, W.: Farbenfabriken Bayer, Germany (1972).

HOFFMAN, H.: Farbenfabriken Bayer AG, Leverkusen, Germany (1972).

HOMEYER, B.: Nemacur, a highly effective nematicide for protective and curative application Pflanzensch. Nach. **24**, 48 (1971).

—— Farbenfabriken Bayer AG, Leverkusen, Germany (1972).

STECHER, P. G. (ed.): The Merck Index, 7 ed. Rahway, N. J.: Merck (1960).

THORTON, J. S.: Determination of residues of Nemacur and its metabolites in plant and animal tissues. J. Agr. Food Chem. **19**, 890 (1971).

WAGGONER, T. B.: Metabolism of Nemacur [ethyl 4-(methylthio)-*m*-tolyl isopropylphosphoramidate] and identification of two metabolites in plants. J. Agr. Food Chem. **20**, 157 (1972).

ZECK, W. M.: A rating scheme for field evaluation of root-knot nematode infestations. Pflanzensch. Nach. **24**, 141 (1971).

—— Private communication (1972).

Manuscript received September 4, 1973; accepted October 17, 1973.

New aspects of organophosphorus pesticides. VIII. Structure and fungitoxicity of organophosphorus compounds.

By

HENRY TOLKMITH* and DORSEY R. MUSSELL*

Contents

I. Introduction

Antifungal agents are used for a variety of purposes of several kinds. There is the necessity for prevention and treatment of mycotic diseases of man, of his animals, and of economically important plants. This may also require sterilization of part of the environment, for instance of soil or air. Furthermore, there are problems involving the preservation of valuable matter, such as foods, fibers, and certain other products.

Mycoses of economically important plants, particularly their prevention and treatment, are usually viewed as a pesticidal matter but can also be considered in a perspective parallel to medical and veterinary mycology. Since the phylogenetic differences between plant hosts and fungal pathogens are smaller than those between mammalian hosts and such pathogens it is not surprising that mycoses are of great importance as plant diseases. As a matter of fact, there are some 8,000 fungal species that are pathogenic for plants, while the number of fungal pathogens for man and domesticated animals is less than 100.

* Dow Chemical U.S.A., Midland, Michigan 48640.

Table I. *Plant mycoses of major importance.*

Disease	Host	Pathogen	
		Species	Class
Pythium seed rot	Various plant seeds	*Pythium* sp.	Phyco-
Late blight	Potatoes	*Phytophthora infestans*	mycetae
Downy mildew	Grapes	*Plasmopara viticola*	
Rhizopus soft rot	Fruits, vegetables	*Rhizopus* sp.	
Powdery mildew	Apples	*Podosphaera leucotricha*	Asco-
	Peaches, roses	*Sphaerotheca pannosa*	mycetae
Dutch elm disease	—	*Ceratocystis utmi*	
Apple scab	—	*Venturia inaequatis*	
Brown rot	Stone fruits	*Sclerotinia fructicola*	
Bunt	Wheat	*Tilletia* sp.	Basidio-
Loose smut	Barley, wheat	*Ustilago nuda*	mycetae
Corn smut	—	*Ustilago maydis*	
Stem rust	Wheat	*Puccinia graminis*	
Rice blast	—	*Piricularia oryzae*	Deutero-
Gray mold	Grapes, strawberries	*Botrytis cinerea*	mycetae
Fusarium wilt	Tomatoes	*Fusarium oxysporum*	

II. Mycoses

In view of the great preponderance of fungi pathogenic for plants we may expect the number of economically important fungal diseases of these hosts to be greater than the number of fungal disease of mammals. However, these two kinds of mycoses numerically are rather similar.

It appears noteworthy that important fungal diseases of mammals very predominantly involve pathogens of the class of deuteromycetes but none of the class of basidiomycetes, while important mycoses of plants are caused by pathogens that belong to all four taxonomic classes of fungi. These diseases, their hosts, and their pathogens are listed in Table I.

III. Structural types of antimycotics

In analogy to the comparative pathology of mycoses, antimycotic agents also may be viewed in a comparative perspective. When doing so we notice that those antimycotics which are of medical and veterinary importance are few in number and represent heterocyclic antibiotics containing oxygen in their heterorings, e.g., amphotericin B, griseofulvin, hamycin, and nystatin.

Commercially important organic plant fungicides, in contrast, are numerous, usually not of microbial origin, and comprise nonheterocyclic structures (Table II) as well as heterocyclic structures. The latter group contains one kind of heteroatom (Table III) or two kinds of heteroatom (Table IV).

For some time one of the major goals of antimycotic research has been the development of systemic plant fungicides since economically

Table II. *Nonheterocyclic plant fungicides.*

Acyclic compounds	
dithiocarbamates	dodine
Carbocyclic compounds	
binapacryl	dichlone
chloroneb[a]	dinocap
chlorothalonil	tecnazene
dichlofluanid	thiophanates[a]
Organotin compounds	
fentin acetate	fentin hydroxide

[a] Systemic fungicides.

Table III. *Heterocyclic plant fungicides involving one kind of heteroatom.*

Containing ring-nitrogen	
benomyl[a]	ethirimol[a]
captafol	folpet
captan	glyodin
Cela W524[a] fungicide	parinol
cycloheximide[a]	piperalin
dichlozoline	triarimol[a]
dimethirimol[a]	
Containing ring-oxygen	
antimycin A	kasugamycin
griseofulvin[a]	pyracarbolid[a]
Containing ring-sulfur	
dithianon	

[a] Systemic fungicides.

Table IV. *Heterocyclic plant fungicides involving two kinds of heteroatom.*

Containing cyclic nitrogen plus oxygen	
blasticidin S	polyoxins
drazoxolon	tridemorph[a]
Containing cyclic nitrogen plus sulfur	
oxythioquinox	thiabendazole[a]
Containing cyclic oxygen plus sulfur	
carboxin[a]	oxycarboxin[a]

[a] Systemic fungicides.

these seem more attractive. The systemics known at the present time very predominantly are of heterocyclic structure and—interestingly—include a compound that was first introduced as a promising medical-veterinary anthelmintic (thiabendazole). It is also noteworthy that over one-third of the commercially important foliage fungicides are systemic.

IV. Fungitoxic organophosphorus compounds

Most of the antimycotics mentioned so far were developed during the past four decades, in contrast to the inorganic antifungals of the preceding century. A structurally novel type of plant fungicide, involving organic phosphorus compounds, emerged during the past decade. This kind of pesticide can be divided into two distinctly different classes of structure, as discussed below.

a) Nonheterocyclics

This class includes three types of structure, *viz.*, phosphorothiolates (Table V), phosphonothionates (Table VI), and phosphorothiolothionates (Table VII). Their chemistry has been reviewed by WEGLER (1970). The phosphorothiolates were developed during 1964 to 1968, following the discovery that Kitazin proved to be a good protectant against, and eradicant of, rice blast. Compounds of this type tend to be quite active against this disease and thus compete with antibiotics such as blasticidin S and kasugamycin. Phosphonothionates show a similar picture of selective activity against rice blast. The phosphorothiolothionates, however, possess some activity as soil fungicides and have not been commercialized.

Main characteristics of practically important nonheterocyclic phosphorus-sulfur fungicides may be summarized as follows: presence

Table V. *Fungitoxic phosphorothiolates.*

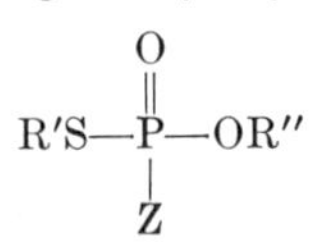

R′	R″	Z	Oral LD_{50} (mg/kg)	Trade name
$(C_6H_5)CH_2$	C_2H_5	OC_2H_5	235 (mouse)	Kitazin
$(C_6H_5)CH_2$	i-C_3H_7	OC_3H_7	660 (mouse)	IBP
$(C_6H_5)CH_2$	n-C_4H_9	SC_2H_5	120 (mouse)	Conen
C_6H_5	C_2H_5	SC_6H_5	218 (mouse)	Hinosan
4—Cl—C_6H_4	CH_3	OC_6H_{11}	160 (rat)	Cerezin

Table VI. *Fungitoxic phosphonothionates.*

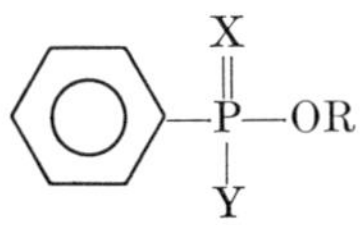

R	X	Y	Oral LD_{50} (mg/kg)	Trade name
CH_3	O	$SCH_2(C_6H_5)$	720 (mouse)	Inezin
CH_3	S	OCH_2NO_2	—	—
CH_3	S	$OC_6H_4CH{=}NNH(C_6H_5)$	—	—

Table VII. *Fungitoxic phosphorothiolothionates.*

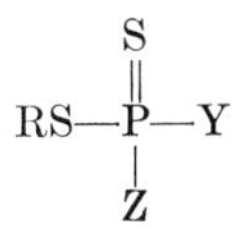

R	Y	Z	Oral LD_{50} (mg/kg)
CH_3	OC_2H_5	OC_2H_5	156 (mouse)
CH_3	SCH_3	SCH_3	—

of a common structural moiety (Fig. 1), a rather narrow spectrum of activity, tendency toward mammalian toxicity, and being acceptable substitutes for organic mercurials in the control of rice blast.

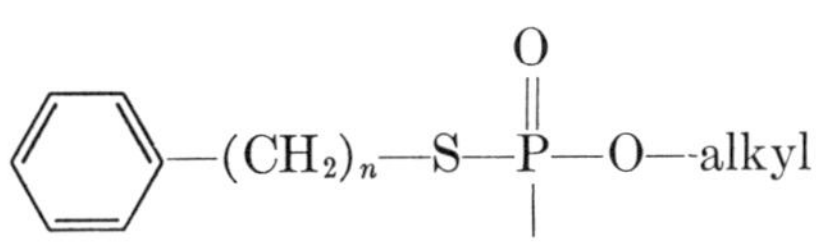

(n = 1 or zero)

Fig. 1. Common structural moiety present in antifungal phosphorothiolates.

b) Heterocyclics

The other class of fungitoxic phosphorus compounds comprises derivatives of nitrogen heterocyclics. In view of the rather diversified nature of the parent structures involved we may consider these compounds in the order of increasing number of N-atoms in their heteroring. Thus, the products to be discussed are Plondrel fungicide, some imidazole compounds, triamphos, and pyrazophos.

Plondrel[1] fungicide (TOLKMITH 1966 b, TOLKMITH and MUSSELL 1967, TOLKMITH *et al.* 1967 d) is an excellent ascomyceticidal eradicant and has the structure of an *O,O*-diethyl phthalimidophosphonothionate (Fig. 2). It controls powdery mildew (*Sphaerotheca, Podosphaera, Erysiphe*) at 0.02 to 0.06 percent of active ingredient while the control of apple scab, leaf spots (*Physalospora, Coccomyces, Diplocarpon*), and brown rot requires somewhat higher concentrations. The product does not show phytotoxicity on various ornamentals, cucurbits, and deciduous fruits. This product and its principal plant

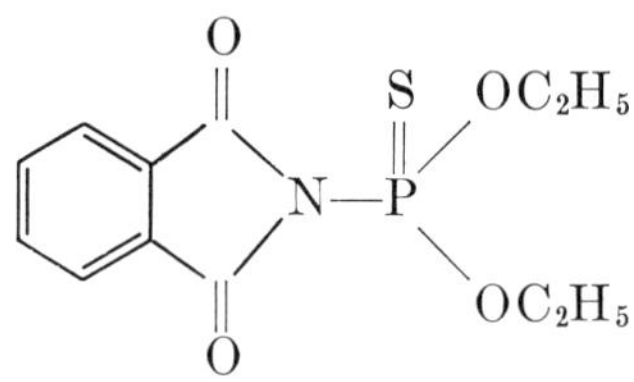

Fig. 2. Structure of Plondrel fungicide.

metabolite, *N*-diethoxyphosphenyl phthalamic acid, have a very low acute mammalian toxicity (TOLKMITH 1966a). There is no indication that this fungicide isomerizes on storage at ambient temperature. In general, the biological properties of Plondrel appear to be related to the fact that it is a weak phosphorylating and alkylating agent but an efficacious acylating agent. In biochemical terms one might say that its C—N bonds are relatively energy-rich, while its P—N bond is not.

Another group of fungicidal phosphorus derivatives of nitrogen heterocyclics, the imidazolylphosphinothionates (Fig. 3), tend to be ascomyceticidal as well as phycomyceticidal (TOLKMITH *et al.* 1967 a and c) and are still under investigation. The degree of activity appears to depend more upon size and shape of the substituents than upon their nature. This observation indicated the possibility that antifungal activity might not substantially change if the tetrahedral phosphorus moiety attached to imidazole were replaced by a moiety of similar shape but of different structure. The hypothesis proved to be

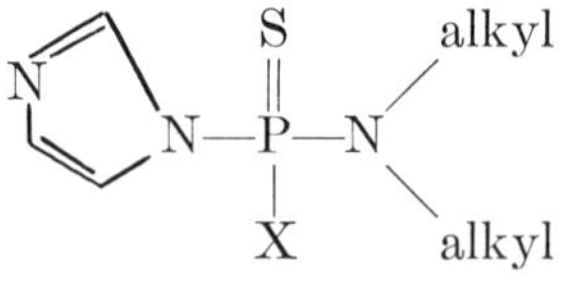

(x = hydrocarbyl, imidazolyl, dialkylamido)

Fig. 3. Composition of fungitoxic imidazolylphosphinothionates.

true with a trityl moiety (TOLKMITH and MUSSELL 1967). Therefore, the antifungal activity of the imidazole derivatives described was evi-

[1] Trademark of The Dow Chemical Company.

dently not a function of their phosphorylation ability, but seemed to involve the nucleophilic imidazole nitrogen. Interestingly enough, a monochlorinated trityl group at imidazole (Fig. 4) was subsequently recognized as useful for the treatment of aspergillosis, candidiasis, and several other systemic mycoses in man.

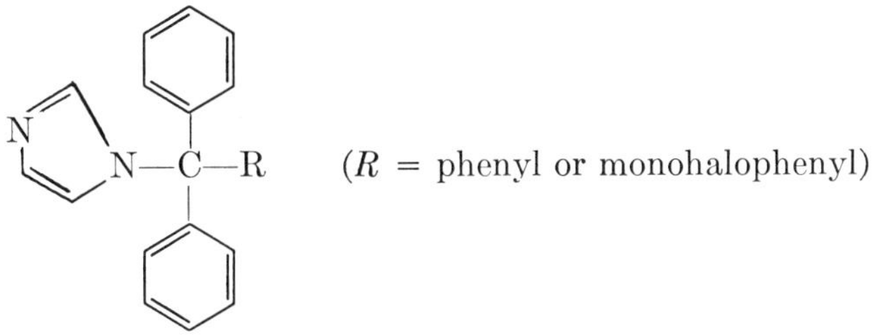

Fig. 4. Constitution of antifungal tritylimidazoles.

Finally, there are two fungicidal phosphorus derivatives of heterocyclics that contain three nitrogen heteroatoms. One is known as triamphos (Fig. 5) and has been originally described by VAN DEN BOS *et al.* (1960). It controls powdery mildew and has systemic insecticidal and acaricidal properties. However, it has a rather high mam-

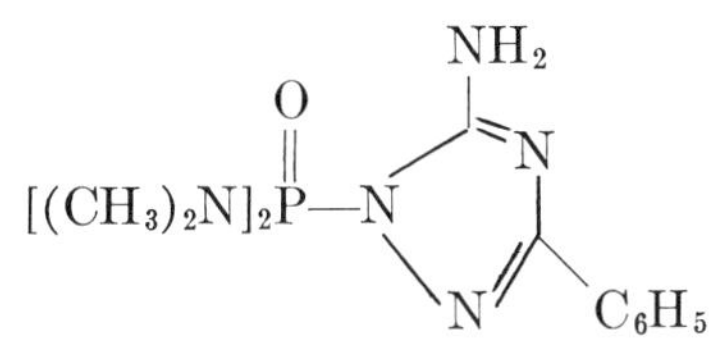

Fig. 5. Structural formula of triamphos or Wepsyn®.

malian toxicity, as indicated by an LD_{50} of 20 mg/kg for male rats and is no longer marketed in the United States. The other product of this group is pyrazophos (MILDENBERGER and SCHERER 1970). Its composition is indicated in Figure 6 and it has an oral LD_{50} of

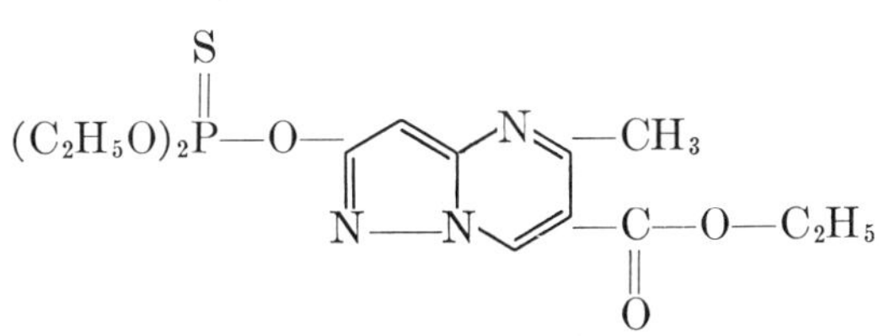

Fig. 6. Structure of pyrazophos.

140 mg/kg for rats. Also known as Afugan[2], it represents a systemic fungicide and is recommended for the control of powdery mildew

[2] Trademark of Farbwerke Hoechst.

on various crops and ornamentals. Like Plondrel® it has a protective and curative action.

Thus, the main characteristics of antifungal phosphorus derivatives of nitrogen heterocyclics differ from those of the nonheterocyclic phosphorus-sulfur fungicides in that they have a different and broader spectrum of activity, show a considerable variation in mammalian toxicity, and do not have a common structural moiety.

c) *Asymmetric structures*

A review of fungitoxic organophosphorous compounds would not be complete without a discussion of the effects of an optically active, *i.e.*, structurally asymmetric phosphorus atom, on biological activity. Several compounds described here contain such an atom and some of these are derivatives of unsymmetric *O*-alkyl phosphorothiolic acids. These acids are racemic and could be resolved by first reacting them with optically active bases and by reacting the alkali metal salts of the isolated optical isomers with benzylchloride. This well-worn method, however, has apparently not so far been applied to the resolution of compounds such as Conen, Cerezin (Table V), and Inezin (Table VI). In contrast, the resolution of asymmetric imidazolylphosphinothionates (Fig. 3) has been accomplished by a novel method of resolution. The essential feature of this unique method, which holds promise for the preparation of a great variety of optically active compounds containing phosphorus as their asymmetric center, is a novel dimethylimidazolium cation (Fig. 7). This moiety is capable

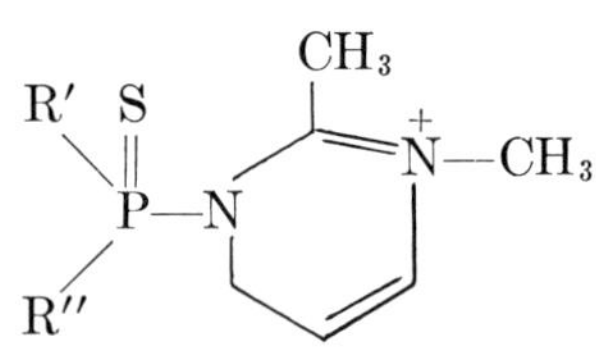

Fig. 7. Constitution of a new dimethylimidazolium cation.

of serving two different purposes, *viz.*, combined with an optically active anion it permits resolution of the racemic base involved and—as a facile leaving group—it is readily displaced on phosphorus by various nucleophiles (SEIBER and TOLKMITH 1967).

On this basis we have been able to determine the biological activity of optically-active imidazolylphosphinothionates, with results as shown in Table VIII (TOLKMITH and MUSSELL 1967, TOLKMITH *et al.* 1967 b). We see that the optical isomers involved do not differ in their antifungal activity but do differ in their mammalian toxicity. As far as the latter is concerned, the *l*-isomer is more toxic than the *d*-isomer. This apparently is the first phosphorothionate, *i.e.*, the

Table VIII. *Optical and biological activity of an imidazolylphosphinothionate.*

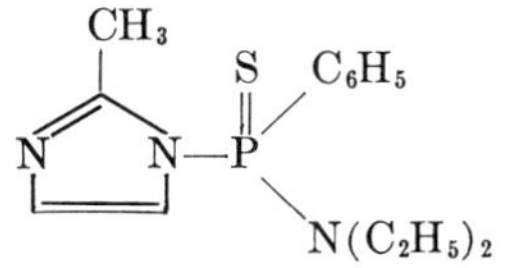

Compound	Optical activity in $CHCl_3$ $[a]_D^{25}$	Control (ppm) of *Phytophthora infestans*	Oral LD_{50} (mg/kg, mice)
Racemate	—	~200	147
d-Isomer	+7.9	~200	316
l-Isomer	−9.3	~200	147

first *indirect* cholinesterase inhibitor, with which a difference in the mammalian toxicity between its optical isomers has been demonstrated (Tolkmith *et al.* 1967 b). Moreover, the differences in toxicity reported with the enantiomers in Table VIII are opposite those known with tabun. This compound is one of the most potent direct anticholinesterases and its *d*-isomer is considered markedly more toxic than its *l*-isomer.

Summary

Practically important mycoses of plants, unlike those of mammals, are caused by pathogens that belong to all four taxonomic classes of fungi. Commercially important organic plant fungicides are usually not of microbial or plant origin and comprise nonheterocyclic as well as heterocyclic structures. The systemic plant fungicides known at the present time very predominantly are heterocyclics.

Fungitoxic organophosphorus compounds have been developed during the past decade and form two distinctly different classes, *viz.*, nonheterocyclic phosphorus-sulfur compounds and phosphorus derivatives of nitrogen heterocyclics. The former have a rather narrow spectrum of activity and are acceptable substitutes for organic mercurials in the control of rice blast, but tend to be rather toxic to mammals and have a common structural moiety.

Members of the other class of phosphorus compound show a different and broader spectrum of antifungal activity, a considerable variation in mammalian toxicity, and do not possess a common structural moiety. Even the presence of the phosphorothionyl group is not of critical importance. Also, there are good reasons for believing that at least some of the phosphorus derivatives of nitrogen heterocyclics act in fungi by a mechanism that is not based upon phosphorylation.

References

MILDENBERGER, H., and O. SCHERER: Pesticidal compositions containing 2-(O,O-diethylthionophosphoryl) pyrazolopyrimidines. German Pat. 1,642,262 (1970).

SEIBER, J. N., and H. TOLKMITH: Resolution of asymmetric organophosphoramides and their stereospecific transformations. Tetrahedron Let. **34**, 3333 (1967).

TOLKMITH, H.: Acute mammalian toxicity and structure of heterocyclic organophosphorus compounds. Ann. N.Y. Acad. Sci. **136**, 59 (1966 a).

—— Biological activity of phosphoramidothionates. Nature **211**, 522 (1966 b).

——, and D. R. MUSSELL: Novel *N*-heterocyclic fungicides. World Rev. Pest Control **6**, 74 (1967).

——, P. B. BUDDE, D. R. MUSSELL, and R. A. NYQUIST: Imidazolyl phosphinamidothionates. J. Med. Chem. **10**, 1074 (1967 a).

——, D. W. OSBORNE, and E. H. BLAIR: Advances in the field of phosphoramidates. 153rd Nat. Meeting Amer. Chem. Soc., Miami Beach (1967 b).

——, J. N. SEIBER, P. B. BUDDE, and D. R. MUSSELL: Imidazole: Fungitoxic derivatives. Science **158**, 1462 (1967 c).

——, H. O. SENKBELL, and D. R. MUSSELL: Fungicidal phthalimidophosphonothionates. Science **155**, 85 (1967 d).

VAN DEN BOS, B. G., M. J. KOOPMANS, and H. O. HUISMAN: Investigations on pesticidal phosphorus compounds. I. Fungicides, insecticides, and acaricides derived from 3-amino-1,2,4-triazole. Rec. Trav. Chim. **79**, 807 (1960).

WEGLER, R. (ed): Chemie der Pflanzenschutz- und Schädlingsbekämpfungsmittel. Berlin-Heidelberg-New York: Springer-Verlag (1970).

Manuscript received September 4, 1973; accepted October 17, 1973.

Toxicological and residue data useful in the environmental safety evaluation of dalapon*

By

EUGENE E. KENAGA**

Contents

* DALAPON is the trademark of The Dow Chemical Company abroad, while "dalapon" is the common name approved by the British Standards Institute and the Weed Science Society of America.

** The Dow Chemical Company, Midland, Michigan 48640.

I. Introduction

The complete environmental evaluation of any pesticide or other chemical involves consideration of the dosages applied, the movement and fate of residues of the compound and degradates in soil, water, air, plants, and animals, and the relationship of these residues with toxicity of living organisms. This review involves such an evaluation of dalapon (2,2-dichloropropionic acid) and its sodium and magnesium salts.

II. Registered uses and food tolerances

Dalapon formulations are registered in the United States for selective use in controlling grasses in crops including apples, apricots, asparagus, bananas, beans, birdsfoot trefoil, citrus (grapefruit, limes, oranges, tangerines), coffee, corn (field and sweet), cotton, cranberries, flax, grapes, Macadamia nuts, peaches, pears, peas, pineapples, plums, potatoes, sorghum, soybeans, sugar beets, and sugarcane [40 CFR (Code of Federal Regulations) 180.150].

Tolerances which have been established by the Environmental Protection Agency for dalapon on these crops or their seeds vary from 0.1 to 75 ppm. These tolerances include a 0.1 ppm residue for milk and 0.2 ppm for meat and meat by-products of cattle, goats, hogs, and sheep and 0.3 ppm residue for eggs, and three ppm for poultry (except for nine ppm for kidney of poultry) based on residues found in or on animals fed dalapon. Dosages of dalapon used for the above purposes vary from 0.74 to 15 lb./A (= 0.85 to 17 lb. of dalapon sodium salt/A or one to 20 lb. of DOWPON[1] or DOWPON M grass killers/A, most applications being less than seven lb./A.

In addition, dalapon is registered for the control of grasses and cattails in noncrop areas such as lawns, railroad ballast and berm areas, industrial areas, drainage ditches, and other nonagricultural lands). Recommended dosages of dalapon for control purposes range from 11 to 22 lb./A (= 15 to 30 lb. of DOWPON or DOWPON M grass killers for most uses.

III. Identity

The Dow Chemical Company sells dalapon formulated as the sodium and magnesium salts. Dalapon in these salt forms is considered equivalent on an ionic or acid basis and also from the standpoint of biological activity.

DOWPON M grass killer is the present major product containing dalapon and consists of dalapon sodium salt 72.5 percent, dalapon magnesium salt 12.0 percent, and inert ingredients 15.5 percent (74 percent dalapon). DOWPON grass killer is an older product and consists of dalapon sodium salt 85 percent and inert ingredients 15 percent (74 percent dalapon). Technical dalapon sodium salt, often in the form of DOWPON grass killer, was the compound used to obtain much of the residue and toxicological data reported here from the literature.

Common names: dalapon, dalapon sodium salt, and dalapon magnesium salt.

Chemical names: 2,2-dichloropropionic acid, 2,2-dichloropropionic acid sodium salt, and 2,2-dichloropropionic acid magnesium salt.

Structural formula and molecular formula:

dalapon:	$CH_3—CCl_2—COOH$	$C_3H_4Cl_2O_2$
dalapon sodium salt:	$CH_3—CCl_2—COONa$	$C_3H_3Cl_2NaO_2$
dalapon magnesium salt:	$(CH_3—CCl_2—COO)_2Mg$	$C_6H_6Cl_4MgO_4$

IV. Chemical and physical properties

a) Physical constants

Important physical constants are listed in Table I.

b) Solubility

Solubilities in several solvents are listed in Table II.

[1] Trademark of The Dow Chemical Company; see "Identity" section.

Table I. *Some physical constants of dalapon preparations.*

Constant[a]	Dalapon	Dalapon sodium salt	Dalapon magnesium salt
Molecular weight	143	165	308.5
Physical state	Liquid	Solid	Solid
Color	Colorless	White	White-tan
Density	1.389 (22.8°C)	—	62.5 lb./cu. ft.
M.P.	—	Dec. before melting at 193°–197°C	Dec. before melting at 193°C
B.P.	185°–190°C	—	—

[a] Data of The Dow Chemical Company.

Table II. *Solubilities of dalapon preparations.*

Solvent	Solubility (g./100 g. solvent at 25°C)[a]		
	Dalapon	Dalapon sodium salt	Dalapon magnesium salt
Water	>80	90	>80
Acetone	—	0.14[b]	—
Methanol	18.5	17.9[b]	35.9
Ethanol	82.7	<18.5	25.5
Benzene	—	0.002[b]	4.4
Ether	Sol.	0.016[b]	—
Kerosene	—	—	0.055

[a] Data of The Dow Chemical Company.
[b] Herbicide Handbook (1970).

c) Chemical and physical properties in water

Some of these properties are listed below.

1. Partitioning coefficient. — Dalapon in *n*-octanol/water is 5.7/1.0 and dalapon sodium salt in *n*-octanol/water is <1/1.0.

2. Specific gravity (25°/25°C) and pH of water solutions. — See Table III.

3. Ionization constant. — $K_a = 2.94 \times 10^{-2}$

4. pK_a. — This constant has a 1.74 average at 0.001 to 0.1 molarity of acid (The Dow Chemical Company) and 1.84 at 0.05 ionic strength (NELSON and FAUST 1969).

Table III. *Specific gravities of dalapon solutions.*

Percent solution dalapon sodium salt	Sp. gr.	pH
1	1.0049	5.7
10	1.0057	5.5
20	1.1145	5.6
50	1.2978	6.0

5. Hydrolysis. — Hydrolysis of dalapon sodium salt occurs according to the following pathway, leading to the formation of pyruvic acid:

$$CH_3CCl_2CO_2^{\ominus} + Na^{\oplus} + H_2O \rightarrow CH_3COCO_2^{\ominus} + H^{\oplus} + NaCl + HCl$$

This equation is based upon the following two observations: (1) SMITH *et al.* (1957) isolated sodium pyruvate in essentially quantitative yield after hydrolyzing a dalapon solution containing excess sodium bicarbonate at 110°C for 15 minutes and (2) BRUST (1953) observed that two moles of chloride ion were liberated from a one percent dalapon sodium salt solution when hydrolyzed at 100°C for 15 hours.

Dalapon sodium salt was hydrolyzed only slowly at 25°C in either dilute or concentrated solution as shown in Table IV. At 50°C, the rate of hydrolysis was more rapid, with approximately 25 percent hydrolysis in eight days at initial concentrations ranging from one to 18 percent in water.

Table IV. *Hydrolysis of dalapon* (BRUST 1953).

Conditions of hydrolysis		Hydrolysis (%)[a]		
		Initial conc. dalapon sodium salt in water		
Temp. (°C)	Time (hours)	1%	5%	18%
25	102	0.27	0.43	0.53
	175	0.41	0.61	0.80
50	20	2.8	3.3	3.6
	95	16	15	17
100	1.5	53	48	45
	14.5	97	81	70

[a] Percent hydrolysis $= \dfrac{\text{moles chloride ion formed}/2}{\text{moles dalapon sodium salt}} \times 100$

Hydrolysis of solutions of either dalapon or dalapon sodium salt are accelerated at alkaline pH values. For example, at 60°C hydrolysis of dalapon sodium salt was 20 percent complete in 30 hours at which time the equilibrium pH was 2.3. In contrast, hydrolysis was 40 percent complete in 30 hours if a pH of 12 was maintained during the hydrolysis (TACEY and BELLINGER 1958).

Considering all of the data on hydrolysis discussed above, it is reasonable to conclude that both dalapon salts and dalapon[2] have chemical hydrolytic half-lives of several months at temperatures less than 25°C and solution concentrations of less than one percent. This is in contrast to data reviewed later in this review which illustrate that dalapon and dalapon sodium salt are essentially completely degraded by microorganisms in soil or in water in approximately one month under most environmental conditions associated with herbicide use. Thus, from an environmental viewpoint, chemical hydrolysis of the herbicide does not appear to be a particularly significant degradative pathway.

6. Photodegradation. — Irradiation of aqueous solutions of dalapon with a reactor lamp providing 253.7 nm of ultraviolet energy at 49°C resulted in significant rates of photodegradation (TANAKA 1972 b). For example, at an initial concentration of 0.25 molar, 70 percent of the dalapon was photodegraded in seven hours. This rate may be compared with only about one percent hydrolysis of dalapon sodium salt in the dark for seven hours at 50°C (see data tabulation in previous section).

TANAKA (1972 b) demonstrated that the principal photodegradation reactions involve formation of pyruvic acid, which is then decarboxylated to carbon dioxide and acetaldehyde (at a slower rate):

$$CH_3CCl_2CO_2H \xrightarrow{H_2O} CH_3COCO_2H + 2HCl$$
$$CH_3COCO_2H \longrightarrow CO_2 + CH_3CHO$$

---

[2] Because of ionization equilibria, the hydrolysis of dalapon and dalapon sodium salt will be identical at a given pH and, except at pH values below 2 will most likely involve hydrolysis of the ionized form which is predominant. In unbuffered systems the solution will become more acidic as the hydrolysis proceeds:

$$CH_3CCl_2CO_2^{\ominus} + H_2O \rightarrow CH_3COCO_2^{\ominus} + 2H^{\oplus} + 2Cl^{\ominus}$$

In nature it is highly doubtful that solutions of less than pH 2 would be encountered. Further, the concentration of dalapon or dalapon sodium salt found in the environment would not be sufficient to alter the pH significantly regardless of which form was added. From the tabulated data, one observes that the rate of hydrolysis is slower in concentrated solution than in dilute solution (at 100° but not at 25°C). This apparent concentration effect is more likely a pH effect, since the solution becomes quite acidic at the high rates of conversion at 100°C (TACEY 1958). TANAKA (1972 a) similarly studied the hydrolysis of solutions of dalapon sodium and observed that the pH's of solutions of the herbicide decrease with time.

Approximately two to four percent of the dalapon undergoing degradation was converted to 1,1-dichloroethane, presumably by direct decarboxylation. Also, small amounts of a water-insoluble polymer were isolated; characterization studies indicated it was derived from the degradates pyruvic acid and acetaldehyde.

The absorption spectra of aqueous dalapon solutions show weaker absorption in the range of 290 to 350 nm than at 254 nm. Thus, under environmental conditions, photodegradation of dalapon may be expected to be slower than quantitized under the conditions of the above study. TANAKA (1972 b), however, showed that oxygenation of solutions of dalapon had a modest positive effect on the rate of photodegradation.

TINDLE *et al.* (1970) showed that ^{14}C-dalapon at an initial concentration of 1000 ppm was 60 percent converted to pyruvic acid after standing one month under normal laboratory lighting conditions.

V. Metabolic and degradative products and residues in mammals and birds

When fed to animals, dalapon was quickly excreted in the urine (LEASURE 1964). Dogs, fed a single dose of 500 mg. of dalapon sodium salt/kg. of body weight excreted approximately 65 to 70 percent of this amount in two days. In repeated daily dietary feeding tests equivalent to doses of 50 and 100 mg. of dalapon sodium salt/kg./day over a 60-day period the excretion of dalapon in the urine ranged from 25 to 53 percent of the total ingested weight of dalapon. The half-life of dalapon in blood appeared to be approximately 12 hours (HOERGER 1969).

Human subjects consuming five successive daily oral doses of 0.5 mg. of dalapon sodium salt/kg. (198.3 mg./person) excreted approximately one-half of the administered dose in the urine over an 18-day period (HOERGER 1969).

REDEMANN and HAMAKER (1959) reported that only two labeled compounds were found in milk from a cow whose feed contained ^{36}Cl-labeled dalapon; these were dalapon and the chloride ion. KUTSCHINSKI (1969) studied the distribution of dalapon residues in milk from dairy cows fed dalapon sodium salt for eight weeks at concentrations of 20 to 300 ppm of dalapon in their dry feed intake. Cows fed 200 ppm of dalapon in their diet produced milk containing 0.7 to 1.0 ppm of dalapon within three to five days. These residue levels did not change appreciably throughout the eight-week feeding period. No significant amount of dalapon was found in fat of milk. The glyceride of dalapon was searched for, but not found (<0.1 ppm).

PAYNTER *et al.* (1960) analyzed kidney, liver, muscle, brain, fat, and milk tissues of rats maintained for two years on diets containing

Table V. *Residue data from rats maintained for two years on diets containing dalapon sodium salt and from milk from lactating female rats* (PAYNTER *et al.* 1960).

Organ or tissues	Dalapon residue[a] (ppm) at dosages in diet of		
	100 ppm	300 ppm	1,000 ppm
Kidney	4.0	9.7	28.2
Liver	1.0	3.5	10.7
Muscle	—	2.9	7.5
Brain	—	2.6	5.4
Fat	—	0.5	1.4
Milk[b]	—	6.0	19.1

[a] Average net values after correction for blanks and recovery.
[b] From third-generation lactating female rats. Milk samples for the rats maintained on diets containing 3,000 ppm of dalapon contained 29.3 ppm of dalapon.

100, 300, and 1,000 ppm of dalapon sodium salt. Residues of dalapon in the tissues were approximately proportional to the dose fed and were at most only a small fraction of the level in the diet (Table V). At the 1,000 ppm level of feeding the highest residues of dalapon were found in the kidney (28 ppm), milk (19 ppm), and liver (11 ppm). As shown in other studies, dalapon does not concentrate in fat.

PAYNTER *et al.* (1960) obtained residue data similar to those on rats by using dogs given daily doses, by capsule, of dalapon sodium salt at the rates of 15, 50, and 100 mg./kg./day, five days a week for one year. This is the equivalent of a high daily dietary consumption of 500, 1,750, and 3,500 ppm. Results of these tests showed residues in all tissues to be quite low in relation to the levels administered (Table VI).

Swine consuming a diet containing dalapon sodium salt at levels of 100 ppm (= 4.7 mg./kg./day) and 200 ppm (= 8.6 mg./kg./day) of dalapon for 28 and 49 days were slaughtered without withdrawal from the feed. Greatest levels of residues of dalapon, averaging 6.9 and 10.4 ppm at the two feeding levels, respectively, were found in kidney, with lesser amounts in liver of 4.4 and 6.8 ppm in lean meat of 3.6 and 5.8 ppm and in fat of 2.2 and 2.4 ppm. No increase in residue was attributed to extending the time of feeding from 28 to 49 days (GETZENDANER 1969 a).

Steers were fed a diet containing dalapon at levels of ten, 30, and 100 ppm, for a period of four weeks. Three animals at each

Table VI. *Residue data from dogs given daily doses (five days a week) by capsule of dalapon sodium salt for one year* (PAYNTER *et al.* 1960).

Organ or tissues	Dalapon residue[a] (ppm) at dosage[b] of		
	15 mg./kg./day	50 mg./kg./day	100 mg./kg./day
Kidney	20	26	69
Liver	22	21	52
Muscle	13	20	43
Brain	9	11	21
Fat	1	3	10

[a] Average ppm corrected for blank and recovery.
[b] Equivalent to about 500, 1,750, or 3,500 ppm in diet.

level were slaughtered without a withdrawal period. Three animals, after four weeks on the 100 ppm level, were fed four days on untreated feed before slaughter. Tissues analyzed and maximum residues found are shown in Table VII. Residues of dalapon in blood, fat, kidney, liver, and muscle tissues remained far below those in the diet. When the chemical was removed from the diet, tissue residues in animals fed 100 ppm of dalapon decreased at a half-value rate of about two days (GETZENDANER 1969 b).

GETZENDANER (1968) reported residue data using laying White Rock hens fed for 30 to 60 days before sacrifice on diets containing

Table VII. *Residue data from steers fed four weeks on diets containing dalapon* (GETZENDANER 1969 b).

Organ or tissue	Dalapon residue[a] (ppm) at dosage in diet of			
	10[b] ppm	30[b] ppm	100[b] ppm	100[c] ppm
Blood	0.2	1.3	5.9	1.3
Fat	0.03	0.3	1.1	0.4
Kidney	0.0	0.4	2.5	0.3
Liver	0.0	0.2	1.7	0.1
Muscle	<0.03	0.3	1.1	0.2

[a] Average values.
[b] Zero days withdrawal from dalapon.
[c] Four days withdrawal from dalapon.

Table VIII. *Residue data from chickens fed 30 days on diets containing dalapon* (GETZENDANER 1968).

Organ or tissues	Dalapon residue[a] (ppm) at dosage in diet of	
	25 ppm	50 ppm
Egg[a]	0.7	1.4
Lean Meat	3	5
Skin, Fat	4	9
Liver	6	12
Gizzard	4	9
Kidney	18	22
Blood	11	22
Feathers	2	4

[a] Average values.

25 and 50 ppm of dalapon. Lean meat, skin and fat, liver, gizzard, kidney, blood, and feathers were analyzed for residues and averaged for the 30-day feeding study as shown in Table VIII. Residues of dalapon in hens' eggs appeared to reach a fairly constant level after about two weeks of feeding and never exceeded two ppm. For the other tissues average residues for birds fed 50 ppm were highest for blood and kidney tissues (22 ppm); others were considerably less. Residues from hens fed 25 ppm were roughly half those at 50 ppm of dalapon, and 30-day residue values were not significantly different from 60-day values for each concentration tested.

VI. Bioconcentration studies in aquatic animals and plants

The main purpose of bioconcentration studies is to determine the distribution of residues, selective or otherwise, of a given chemical over time in different organisms and segments of their environmental surroundings which constitute typical or representative ecosystems. It is important to determine whether residues soon reach a maximum concentration or whether residues increase over periods of days, weeks, or months. In many studies the residues are often not measured before the application of the chemical to establish background levels from unknown sources, nor immediately afterwards. Because of this deficiency, two sequential phases of residue bioconcentration are not differentiated: (1) physical adsorption to surfaces which occurs almost immediately and which is related to surface area/volume measurements of the organism and (2) long-term organism penetration which is absorption. This differentiation is important since continued build-up of residues in organisms over time is mostly due to absorption and redistribution within the organism and is a function of its metabolism. Adsorption factors may result in residues which constitute a base-line of residues to be subtracted from total (absorbed and adsorbed) residues, thus helping to determine those residues due to penetration of the chemical into the organism over a period of time (bioconcentration because of absorption).

SMITH *et al.* (1972) studied the distribution of dalapon sodium salt, in an aquarium containing five ppb of dalapon, in the water, fish, snails, *Daphnia* sp., submerged plants (*Cabomba caroliniana*), floating plants (*Limna minor*), and algae over a three-day exposure period. A second identically furnished aquarium was used except that instead of using live organisms, the plants were killed by autoclaving and the animals were killed by freezing them in Dry Ice prior to putting them in the aquarium. The differences with time between residues from organisms in the two aquariums were essentially that due to adsorption and absorption in the first aquarium minus that due to adsorption in the second aquarium. Concentrations of dalapon in water remained at five to six ppb throughout the exposures, showing little or no volatility from water or overall adsorption from water by organisms. Fish appeared to contain up to 15 ppb on or in tissues of both dead and live whole fish. Snails and *Daphnia* sp. contained residues less than that of the surrounding water environment. Duckweed (*Limna minor*) contained about 40 to 50 ppb when dead and up to about 180 ppb when alive. *Cabomba caroliniana* contained about 20 to 50 ppb when dead, but around 780 ppb when alive and appeared still to be absorbing residues after 72 hours exposure.

VII. Metabolic and degradative products and residues in plants

a) Metabolism

BLANCHARD *et al.* (1960) used 2-^{14}C dalapon sodium salt to measure the distribution of dalapon in corn and soybean plants. Soybeans were treated by root application in nutrient solution containing 32 ppm of dalapon for four days and by foliar application and held for three days postapplication. Corn was treated by root application in a nutrient solution of 2.5 ppm of dalapon for 11 days. Dalapon tended to concentrate in the young tissues as shown by autoradiograms. Extracts of treated plant tissues showed only a single spot (R_f value) on paper chromatograms, corresponding to that of dalapon. Two plausible breakdown products, pyruvic and lactic acids, have different R_f values and were not found. This is evidence that little or no breakdown of dalapon can be detected in plants from three to 11 days after treatment.

SMITH and DYER (1961) found by use of 2-^{14}C dalapon sodium salt, that dalapon was present in cotton plant tissues as a free acid or salt and could be leached from tissues by simple water and that after prolonged extraction there was no measurable chemical binding or interaction with the plant. Cotton plants grown in soil about eight inches deep, treated with 500 ml. of water containing 71 ppm of dalapon (equals approximately five lb./A), and harvested after 12 weeks, contained no pyruvic acid, acetic acid, or 2-chloropropionic acid as ascertained by use of paper chromatography references against known standards. The only compound found was dalapon.

FOY (1961) employed dalapon 2-^{14}C and ^{36}Cl for tracer and metabolic studies in cotton, corn, and wheat. Dalapon was absorbed, translocated, and redistributed and accumulated in plants principally as the intact molecule or dissociable salt thereof, and remained unchanged for long periods, especially in dormant or quiescent tissues. Slow metabolic decomposition resulted in some release of ^{36}Cl or incorporation of ^{14}C into other compounds indicating eventual breakdown, possibly an initial dehalogenation followed by modified propionate oxidation. Slowness of metabolism or the stability of dalapon may also be indicated by the presence of "dalapon stimulus" in the third generation of seeds after exposure of the first generation of wheat to pre-plant applications of four lb. of dalapon/A in the field. This was detected by typical plant formative effects of dalapon but not by chemical analysis. The degradation of dalapon in plants has been reviewed thoroughly by FOY (1969).

b) *Residues*

FERTIG and SCHREIBER (1961) reported that "early residue studies" on forage legumes, alfalfa, and birds'-foot trefoil showed a maximum of approximately 300 ppm of dalapon by use of a spring application of ten lb. of dalapon/A to these crops.

GETZENDANER (1972) reported analysis of residues from DOWPON grass killer at eight lb./A (equals six lb. of dalapon/A) applied to grass at intervals of zero, one, two, four, six, and eight weeks after application. Average residues of dalapon per one lb. of DOWPON/A after the above time intervals were 135, 28, 25, 21, six, and four ppm, respectively. In terms of average residues of dalapon per one lb. of dalapon/A after the same time intervals they were 182, 38, 34, 28, eight, and five ppm, respectively. These and similar residues from GENTRY (1971) for three other grass analyses are shown in Table IX. Residue data for treatment of asparagus (GETZENDANER 1964) are shown in Table X and are much lower than that of grass.

The amounts of residues of a pesticide on plants are usually related

Table IX. *Residue of dalapon in grass from treatments with* DOWPON *grass killer at eight lb./A.*[a]

Grass type	Residue[a] (ppm) of dalapon/1 lb. of DOWPON/A after							Reference
	0 weeks	1 weeks	2 weeks	4 weeks	6 weeks	8 weeks	10 weeks	
"Michigan"	135	28	25	21	6	4	—	GETZENDANER (1972)[b]
Bahia	100	53	42	28	21	13	—	GENTRY (1971)[c]
Pangola	48	44	21	8	2	1	<1	GENTRY (1971)[c]
Bermuda	57	31	18	12	2	1.5	<1	GENTRY (1971)[c]

[a] Actual residue from application at eight lb./A divided by eight.
[b] Used DOWPON M grass killer, a formulation containing 74 percent dalapon (72.5 percent dalapon sodium salt, and 12 percent dalapon magnesium salt).
[c] Used DOWPON grass killer, a formulation containing 74 percent dalapon (85 percent dalapon sodium salt).

Table X. *Residues of dalapon in asparagus treated during asparagus-cutting time* (GETZENDANER 1964).

Application (lb./A)		Days after application	Dalapon (ppm)[a]	Dalapon (ppm) per 1 lb./dalapon/A
DOWPON	Dalapon			
40	29.6	3	36	1.22
	29.6	4	12	0.41
	29.6	7	6	0.20
20	14.8	3	11	0.74
	14.8	4	8	0.54
	14.8	7	3	0.20
20	14.8	15	1	0.07

[a] Average.

directly to the dosage applied. The residues of dalapon expected on grass and asparagus at various intervals after application can be based on the data in Tables IX and X. With usage dosages of one, 15, and 30 lb. of dalapon sodium salt/A these residues are extrapolated as shown in Table XI and are typical of maximum residues on plant foliage (grass) and stems (asparagus). As discussed by HOERGER and KENAGA (1972) residues are, in part, a function of plant surface area-to-weight ratios, those with the largest ratio such as foliage having the highest residues. Thus, it would be expected that residues on seeds, stems, fruits, etc., would be lower than for foliage with the same dosage application.

Table XI. *Environmental dosage-residue conversion table: Maximum residues in water and soil, and on or in plant tissues, expected as related to a range of dosages of dalapon used for commercial weed control.*

Substrate	Residues of dalapon (ppm)		
	0.74 lb./A (1 lb./A)[d]	11.1 lb./A (15 lb./A)[d]	22.2 lb./A (30 lb./A)[d]
Water[a]			
3 inches deep	1.09	16.4	32.8
1 foot deep	0.27	4.1	8.2
4 feet deep	0.07	1.0	2.0
Soil[a]			
3 inches deep	0.54	8.1	16.2
1 foot deep	0.13	2.0	4.0
Plants			
Grass[b] after			
0 week	48–135	720–2025	1440–4050
1 week	28–53	420–795	840–1790
2 weeks	18–42	270–630	540–1260
4 weeks	8–28	120–420	240–840
6 weeks	2–21	30–315	60–630
8 weeks	1–13	15–195	30–390
10 weeks	<1	—	—
Asparagus[c] after			
0.43 week	0.74–1.2	11–18	22–37
1.0 week	0.20	3	6
2.1 weeks	0.07	1	2

[a] KENAGA (1968), immediately after mixing.
[b] GETZENDANER (1972) and GENTRY (1971), average values.
[c] GETZENDANER (1964).
[d] Equivalent in lb. DOWPON[R] or DOWPON M grass killers/A.

VIII. Metabolic and degradative products and residues in soil

a) Disappearance of residues from soil and microbial degradation by soil organisms

Microorganisms are capable of degrading dalapon and after acclimation in the soil the speed of degradation is greatly increased. THIEGS (1955) found that the concentration of dalapon (59 ppm) in soil sterilized by autoclaving previous to treatment with dalapon did not change after incubation at 100°F for one week while a similar test with unsterilized soil containing microorganisms caused the disappearance of dalapon in four to five weeks after one application and in one week after the second application of 50 ppm.

HOLSTUN and LOOMIS (1956) studied the decomposition of dalapon in soil at 120 ppm by bioassay using elongation of millet seedling shoots for comparative effects of different concentrations of dalapon. Based on known samples, the method was sensitive to about 15 ppm. Dalapon degradation is very sensitive to temperature presumably related to microbial activity. After a two-week incubation period with soil moisture at field capacity, little or no decomposition occurred when the soil temperature was less than 15°C, but decomposition was essentially complete at 30°C. The rate of decomposition was inhibited by low soil moisture, low pH, large additions of organic matter, and temperatures below 20° to 25°C.

JENSEN (1957) found that a group of bacteria (apparently *Agrobacterium*) decomposed dalapon, forming chloride ion in soil after an initial delay of four weeks, probably due to acclimation of the microorganisms; the formation of pyruvic acid was postulated. SWEET *et al.* (1958) and SANFORD and STOVELL (1960) determined that dalapon at about 7.4 to 20 lb./A remained phytotoxic for a month, while SHEETS and HARRIS (1965) reported that dalapon at about 20 lb./A was phytotoxic for three to four months in field tests. MAGEE and COLMER (1969) found eight species of soil bacteria capable of decomposing dalapon. These organisms were capable of overcoming the inhibitory effect of dalapon when grown on agar at 0.5 percent (5,000 ppm). Six of these species were tentatively identified as *Agrobacterium* and two as *Pseudomonas*. Other unidentified actinomycetes, molds, and bacteria were also present.

HIRSCH and ALEXANDER (1960) found that soil cultures of strains of *Nocardia* and *Pseudomonas* microorganisms liberated 90 to 100 percent of the chlorine in dalapon in three weeks. Other genera of bacteria and fungi were referenced as having the ability to utilize *alpha*-chloro-substituted acetic and propionic acids, and included, in addition to those above, *Achromobacter, Trichoderma, Penicillium*, and *Clonostacy*.

DAY *et al.* (1963) reported on the degradation of 50 ppm of dalapon (applied as the sodium salt) in extensive laboratory tests, using 43 California soils. The capacity of the soil to decompose dalapon, as determined analytically, was essentially random with respect to soil series, texture, cation-exchange capacity, total organic matter, and geographical source, but apparently was related to differences in the populations of soil organisms capable of decomposing dalapon. Decomposition of dalapon ranged from complete disappearance in less than two weeks to retention of two-thirds after eight weeks at this high dosage rate (50 ppm = 68.5 lb./A in three-inch soil; see Table XI).

KAUFMAN (1964) studied the microbial degradation of dalapon at six lb./A in five soils varying in pH from 4.1 to 5.0. Some soils were later buffered to pH 7.0. Oat plant bioassays revealed that under greenhouse conditions (28°C) dalapon persisted only four to eight days in muck soil, eight to 16 days in loam and silty loam soils, 16 to 32 days in sandy loam soil, and 32 to 64 days in a silty clay soil. In these three soils, the degradation appeared to be largely a bacterial phenomenon, whereas in the other soils numerous fungi were also present. Each of the five soils contained one or more organisms capable of degrading dalapon. Other factors causing variation in degradation were changes in organic matter, pH, cation-exchange capacity, and aeration. Adaptation of microorganisms capable of degrading dalapon under anaerobic conditions was not observed in any of the five soils investigated. Slight utilization of dalapon was observed under anaerobic conditions with a pure culture of *Arthrobacter* sp. 1.

KEARNEY *et al.* (1964) found that representatives of the following microbial genera were capable of liberating chloride ion from the dalapon ion: *Pseudomonas, Agrobacterium, Bacillus, Alcaligenes, Arthrobacter,* and *Nocardia* by use of pure culture isolation of organisms adapted to dalapon sodium salt. A cell-free preparation of an *Arthrobacter* sp., a purified enzyme system, was found to be capable of dehalogenating dalapon to pyruvate.

BEALL *et al.* (1964) studied the comparative metabolism of 1-^{14}C- and 2-^{14}C-labelled dalapon by soil microorganisms and pure cultures of *Arthrobacter* sp. Microorganisms in muck soil degraded dalapon sodium salt in about two weeks. Subsequent additions disappeared in about two days. ^{14}C from dalapon-1-^{14}C disappeared more rapidly than that from dalapon-2-^{14}C. Much of the 1-^{14}C-labelled material was recovered as $^{14}CO_2$ in two-hour experiments. Some activity appeared in protein and lipid fractions. In contrast, ^{14}C from 2-labelled dalapon was found mostly in the lipid, nucleic acid, protein, and cold-soluble trichloroacetic acid-soluble fractions, and a small amount as $^{14}CO_2$. Under anaerobic conditions with *Arthrobacter* sp. using 2-^{14}C-labelled dalapon, pyruvate and the amino acid alanine were revealed as metabolic products.

CORBIN (1965) determined the influence of pH on the detoxification of dalapon sodium salt in soils by plant bioassay with cucumbers. Dalapon at 2,000 ppm in soil containing *Pseudomonas* sp. was almost 100 percent detoxified within a two-week period at 26°C and pH 6.5, less than 50 percent detoxified at pH 5.3 and pH 7.5 at four weeks, and it was not detectably detoxified at pH 4.3 at eight weeks by the bioassay method. Detoxification of dalapon in liquid culture media was determined by chloride-ion release. There was no degradation within the first two weeks of incubation followed by rapid detoxification of dalapon in the next week at pH 6.5. Dalapon was completely persistent in bacterial cultures buffered at pH 4.5 and 5.5 and only ten percent detoxification occurred at pH 7.5 within three weeks.

BURGE (1969) determined the numbers of bacteria capable of decomposing dalapon sodium salt at 100 ppm in five soils by the most-probable-number method. All soils contained substantial numbers of bacteria sufficient to represent abundant inoculum for rapid utilization of the dalapon substrate if soil conditions were otherwise favorable for the organism to respond. *Arthrobacter* and *Agrobacterium* were found to increase readily on incubation with dalapon, while glucose added to soil increased dalapon-decomposing *Bacillus*. The time required for decomposition of this high dosage varied in the five soils from 30 to greater than 150 days. As an aid to the calculation of dosages of dalapon in various depths of soil, ppm-lb./A factors are given in Table XI.

b) Adsorption and leaching

WARREN (1954) determined the leachability of dalapon sodium salt at eight lb./A in four soil types in comparison with many other herbicides. Dalapon and TCA were the only herbicides to move readily in all soils tested. Bioassays on crabgrass were used to measure activity. HOLSTUN and LOOMIS (1956) studied the leaching of dalapon sodium salt at the rate of 24 lb. a.i./A in glass soil-tube columns by the addition of 1.5 inches of water to the soil column. Successive one-inch deep layers of various types of soils from this column were analyzed for residues by use of a soybean growth bioassay, sensitive to dalapon at one lb./A. Three silty loam and silty clam loam soils were modified by the addition of ten and 20 percent sand, or four and eight percent organic material (manure). None of the soils adsorbed dalapon very much; however, manure decreased and sand increased the downward mobility of dalapon slightly. The bioassay test used involved a phytotoxic response over a ten-to-14 day period and did not necessarily exclude microbial degradation.

SMITH *et al.* (1972) studied the leaching of dalapon-^{14}C at the dosage of one lb./A through an 8.5-inch deep soil column of Kawkawlin sandy loam. Nine acre inches of water applied at one-half inch increments for wetting accounted for the leaching of 99.4 percent of the dalapon in the effluent from the soil column.

HELLING (1971) classified 40 pesticides into five groups based on relative mobilities on silty clay loam plates, using soil thin-layer chromatography based on $\mathbf{R}_f$ values. Only four compounds (ten percent) were classified in the group of greatest mobility (least sorption): TCA, dicamba, amiben, and dalapon. Such materials are among the most likely to leach through or run off soil among the 40 pesticides tested.

The above reports show the potential leachability of dalapon under certain laboratory conditions, often using exaggerated conditions of rainfall or soil permeability. Under actual use conditions MILLER and GETZENDANER (1973) analyzed soils taken from various locations in the United States representing different climates, rainfall, soil types, and soil depths. Samples were taken at intervals of several days to one year after application of DOWPON M or DOWPON grass killers at rates of four to 41 lbs./A. No residues of dalapon were found in the zero-to-six-inch layer of soil from any treatment after 40 days. In a Wayside, Mississippi soil treatment using ten lb. of DOWPON/A the residues of dalapon in the zero-to-six inch layer at seven days were 2.0 to 2.2 ppm, at 14 days 0.11 to 0.28 ppm, at 28 days 0.06 to 0.15 ppm, and at 42 days $<$0.025 ppm. No residues of dalapon were found in the six-to-12-inch layer of soil from any treatment or time interval from application. Dalapon is unstable in soil under conditions favorable for microbial growth. Such field data show that under many practical conditions dalapon does not move beyond the first six-inch depth of soil, probably because microbial action proceeds at a faster rate than leaching under favorable conditions.

IX. Metabolic and degradative products and residues in water

a) Microbial degradation in water

The theoretical Chemical Oxygen Demand (COD) for dalapon sodium salt when completely oxidized to CO_2, H_2O, NaCl, and Cl_2 is 0.53 part of oxygen per part of chemical. The COD test also represents determination of the maximum amount of oxygen required if the chemical were completely consumed for energy purposes by microorganisms (principally bacteria) and is used as a comparison with Biological Oxygen Demand (BOD) data on the same chemical (BOD/COD rates). The COD for dalapon sodium salt, determined by the potassium dichromate method, was 0.53, the same value as the theoretical value for dalapon sodium salt; the COD value for dalapon magnesium salt was 0.57. The BOD test is used to measure the amount of chemical that can be destroyed metabolically through oxidation and used by the life processes of microorganisms.

In order to determine the presence of organisms in soil or aqueous sludges which might be capable of degradating dalapon, three sources were sampled (ALEXANDER and BATCHELDER 1972): an agricultural soil (Kawkawlin sandy loam), a city sewage sludge (Midland), and a chemical waste sludge (The Dow Chemical Company). Each microbial culture was acclimated at 70°F for two weeks to 20 ppm of each chemical. Subsequently, BOD tests were run on acclimated and unacclimated cultures with DOWPON M grass killers and dalapon sodium salt, each at 3.33 and 1.67 ppm. Samples were incubated in the dark at 20°C. All of the microbial cultures used appeared to degrade the chemicals by oxidation; however, the acclimated cultures did so within five days, while unacclimated cultures took more time. Data concerning the Kawkawlin soil cultures are shown in Table XII. The maximum BOD/COD ratios were 67 percent. All of the

Table XII. *BOD measurements with dalapon sodium salt and* DOWPON M *grass killer using aqueous microbial cultures taken from Kawkawlin sandy loam.*

Chemical or formulation	Culture acclimation, exposure time	Parts BOD/part dalapon sodium salt or DOWPON M	BOD/COD[a] (%)
Dalapon sodium salt	*Unacclimated*		
	5 days	0.04	7
	10 days	0.30	56
	20 days	0.34	63
	Acclimated		
	5 days	0.34	63
	10 days	0.36	67
	20 days	0.36	67
DOWPON M	*Unacclimated*		
	5 days	0.04	7
	10 days	0.30	56
	20 days	0.30	56
	Acclimated		
	5 days	0.32	59
	10 days	0.36	67
	20 days	0.36	67

[a] Theoretical COD for dalapon sodium salt = 0.53 and for DOWPON M grass killer = 0.57.

cultures appeared to reach the maximum BOD at the time that 56 to 67 percent BOD/COD ratios were achieved.

Some materials that are considered to be readily biodegradable by microorganism in five-day BOD tests are sucrose, dextrose, glutamic acid, acetone, and ethanol. The BOD/COD ratios of these compounds are, respectively, 51, 50, 53, 69, and 63 percent which are in the same range as the dalapon sodium salt and DOWPON M grass killer BOD/COD ratios. These data show dalapon sodium salt and DOWPON M to be quite easily degraded with acclimated cultures.

In addition to the above degradation of dalapon, the metabolite of dalapon, pyruvic acid, is part of a reversible energy cycle of cells and may be in part excluded from complete oxidation because of the recycling of pyruvic acid (see section on pyruvic acid).

FOY (1961) found that pure dalapon, dissolved in water, became contaminated by microorganisms despite storage in a refrigerator. The changes which occurred in pure aqueous stock solution offered positive proof that some species of fungi, possibly *Alternaria* sp., are capable of using dalapon as a sole substrate to form new substances.

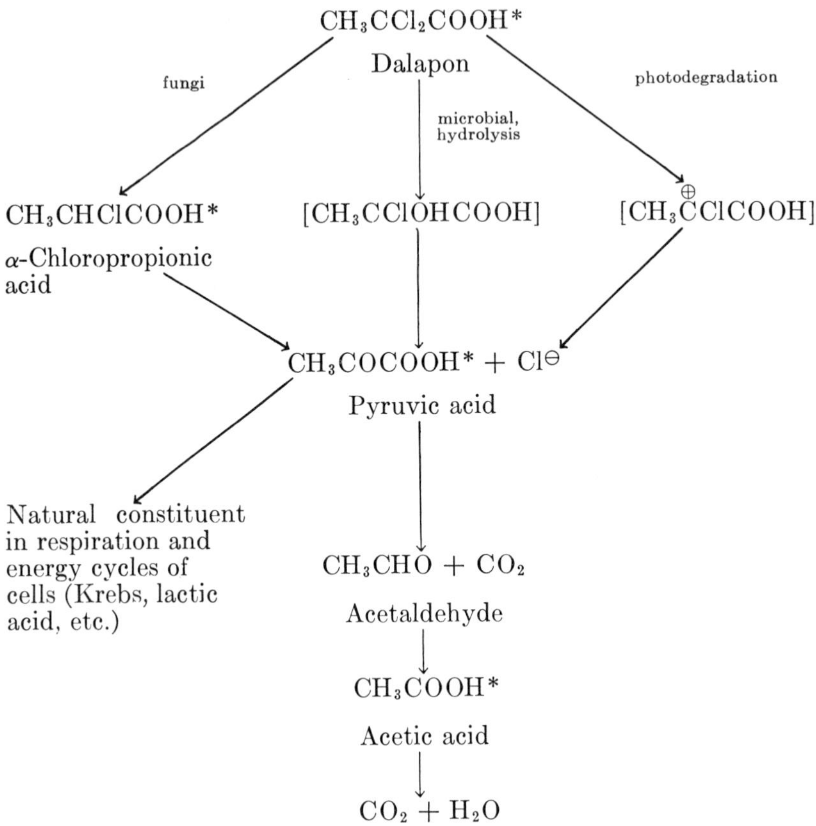

Fig. 1. Chart of the major metabolic and degradative routes of dalapon: the starred acids will be in equilibrium with their anions with the position of the equilibrium depending on pH and the specific cations in a particular environment and the compounds in brackets will be transient.

Among the metabolites formed were inorganic chloride and monochloropropionate (Fig. 1).

b) Residues in irrigation water

Since dalapon is used to control plants which grow on banks of ditches containing flowing water (such as used for irrigation water) there is a need to know the practical hazard to downstream nontarget plants. FRANK *et al.* (1970) studied the concentrations of dalapon in irrigation water following canal bank applications of 6.7 to 20 lb./A. Maximum concentrations of dalapon found were 0.023 to 0.365 ppm when sampled in the treated area within 30 minutes after application. The treated irrigation canals varied in water flow from 3.4 to 290 cu. ft./second and 0.96 to 2.4 ft./second (2.4 ft./second = 3.65 mph). After several hours only very low levels of dalapon (<10 ppb) were present in waters flowing from the treated area. A concentration of 0.365 ppm of dalapon in irrigation water would deposit only 0.09 lb. of herbicide per acre-inch of water and would be too low a rate to be likely to injure irrigated crops.

BRUNS and DAWSON (1959) applied dalapon sodium salt, at the rates of 0.5, 2.5, and 5.0 lb./A in three acre-inches of irrigation water to corn in the five-to-seven leaf stage and rutabagas in the six-to-seven leaf stage. The 2.5 and 5.0 lb./A dosage caused injury to corn, but only slight injury to rutabagas. The 0.5 lb./A dosage caused no significant damage to either corn or rutabagas.

As an aid to the calculation of dosages of dalapon in various depths of water, ppm-lb./A dosage conversion factors are given in Table XI.

X. Some natural cellular functions of pyruvic acid, the major metabolite and degradate of dalapon

Pyruvic acid, a major degradate of dalapon, is also a commonly occurring natural cellular metabolite which occurs in the utilization of carbohydrates for the production of energy in cells. For example, in the presence of the cytochrome series of enzymes and diphosphopyridine nucleotide (DPN), lactic acid is converted to pyruvic acid plus $DPN \cdot H_2$ plus energy. Pyruvic acid is hydrogenated in a reversible system to reform lactic acid and thus the lactic acid-pyruvic acid complex is involved in energy storage or release. Pyruvic acid as formed by glycolysis or other metabolic processes is also the precursor to the citric acid cycle (Krebs cycle) which involves reduction and condensation to form acids containing six carbons. The latter cycle is a series of complicated oxidations of acids involving the production of CO_2 as a function of basic cellular respiration (DILLON 1964). Thus, pyruvic acid, the metabolite and degradate of dalapon, is no different as a molecule from the pyruvic acid needed for energy transfer and respiration at the cellular level in living organisms and can follow the same useful pathways of metabolism and degradation as it does in nature.

XI. Toxicity to mammals

a) *Acute oral tests*

The acute oral LD_{50} values given by PAYNTER *et al.* (1960) and NORRIS (1969) for dalapon sodium salt and dalapon formulations are shown in Table XIII. All of the LD_{50}'s are high in value, indicating very low acute toxicity.

Table XIII. *Acute oral toxicities of dalapon to mammals.*

Species	Sex	LD_{50} (mg./kg.)		
		Dalapon sodium salt[a]	DOWPON[b]	DOWPON M[c]
Rat	M	9,300	5,660	5,660
	F	7,570	5,660	5,660
Mouse	F	>4,600	—	—
Guinea pig	F	3,860	4,930	2,830
Rabbit	F	3,860	2,830	2,140

[a] PAYNTER *et al.* (1960).
[b] Formulation containing 85 percent dalapon sodium salt (NORRIS 1969).
[c] Formulation containing 72.5 percent dalapon sodium salt and 12 percent dalapon magnesium salt (NORRIS 1969).

b) Subacute toxicity tests

1. Oral capsule or drench tests. — PALMER and RADELEFF (1969) dosed cattle and sheep with daily oral capsule or drench applications of dalapon sodium salt at rates of 50 to 500 mg./kg./day, equivalent to an intake of about 150 to 1,500 ppm in the diet; no mortality occurred at any of the dosage rates which were given for from nine to 481 successive days (Table XIV). Cattle exhibited some throat irritation at 500 mg./kg. with an oral drench and sheep lost some weight

Table XIV. *Toxicity of dalapon sodium salt to cattle and sheep* (PALMER and RADELEFF 1969).

No. days treated	Means of oral dosing	Dosage (mg./kg./day)	Effects
Cattle			
10	Drench	250	None
9	Drench	500	Throat irritation
10	Capsule	500	None
Sheep			
10	Drench	50	None
10	Drench	100	None
10	Capsule	100	Ill but survived, 6% wt. loss
481	Capsule	100	None, 10% wt. loss after 86 treatments
10	Drench	250	None
10	Drench	500	Ill but survived, 17% wt. loss
10	Capsule	500	Ill after 7 treatments but survived, 6% wt. loss

at as low as 100 mg./kg./day when fed for ten days. PAYNTER *et al.* (1960) reported the treatment of a heifer and a bull suckling calf with 1,000 mg./kg./day for ten successive days. Some symptoms exhibited were temporary. No mortality occurred. Macroscopic and microscropic examination of tissues from the animals sacrificed on the eleventh day revealed these tissues to be within normal limits, with the possible exception of kidney tissues of the bull.

2. Dietary tests. — PAYNTER *et al.* (1960) conducted a 97-day dietary feeding study on rats using diets containing 115, 346, 1,150, 3,460, and 11,500 ppm. No mortality occurred at any of these dosages. There was no measured effect on male rats at 1,150 ppm (= 115 mg./kg./day) and on female rats at 115 ppm as judged by food consumption, average body and organ weights, and by hematological, gross, and histopathological examinations.

GOLDSTEIN and LONG (1960) conducted field tests with dalapon (sodium salt?) on a grassy pasture using 30 lb./100 gal. of water (= approx. 30 lb./A). Two calves grazed on the pasture until the vegetation turned brown, with no visible harmful effects to the calves. Oats treated with 2.5, five, ten, and 20 lb./100 gal. (= approx. same rates/A) were fed on by calves and sheep without ill effects. Sudan grass treated with 20 lb./100 gal. was fed upon by a calf without ill effect. GOLDSTEIN and LONG (1960) added dalapon sodium salt at the rate of "a pint of the standard concentration of dalapon spray per five gallons of drinking water" (probably equivalent to 12,500 ppm); this was the drinking source for a hog, sheep, and cattle for two weeks without causing visible ill effects.

c) *Chronic tests*

1. One-year dog test. — Dogs were administered capsules daily, five days a week for 52 weeks, containing doses of 15, 50, or 100 mg./kg./day of dalapon sodium salt. The usual extensive toxicological parameters measured indicated no effect from these treatments except increases in average kidney weights at 100 mg./kg./day (PAYNTER *et al.* 1960).

2. Two-year dietary feeding tests with rats. — Rats consumed diets containing 100, 300, and 1,000 ppm of dalapon sodium salt for two years, equivalent to five, 15, and 50 mg. of dalapon sodium salt/kg./day, respectively. The usual extensive toxicological parameters measured during chronic feeding tests indicated no effect from these treatments except increases in average kidney weights at 50 mg./kg./day (PAYNTER *et al.* 1960).

d) *Reproduction tests*

Young rats were fed for at least 110 days on diets containing 300, 1,000, or 3,000 ppm of dalapon sodium salt (= about 30, 100, or

300 mg. of dalapon sodium salt/kg./day) continuously through each of three generations of two litters each, with no effect on reproduction (fertility, gestation, viability) and lactation. No adverse effect was seen on body weight, a favorable indicator of normal growth and maturation (PAYNTER *et al.* 1960).

Dalapon was administered daily (THOMPSON *et al.* 1971) by gavage to pregnant rats from day six through 15 of the gestation in a tolerance study designed to suggest dose levels acceptable for teratogenic studies. Dosages given were 250, 500, 1,000, 1,500, or 2,000 mg./kg./day. The following parameters were examined: maternal body weights and gains, food consumption, number of viable fetuses, fetus resorptions, and corpora lutea, individual pup weights and sex, and gross external examination of pups. As the result of these tests only the following notable observations were made: Clinical signs of toxicity consisted of soft stools and slight appetite depression in dams from the 2,000 mg./kg. dosage group. One rat from the 1,500 mg./kg. group had diarrhea. Weight gains for pregnant dams were lower than controls at 1,500 or 2,000 mg./kg./day. Fetal resorption rate was increased at both the 1,500 and 2,000 mg./kg. dosage levels, but the increase was not significantly different from controls. Pup weights from the 2,000 mg./kg. dosage level were significantly lower than controls. No effects were seen at 1,000, 500, or 250 mg./kg./day.

e) *Teratological studies in rats treated orally with dalapon during organogenesis*

Teratological studies of dalapon in the rat were reported by EMERSON *et al.* (1971). Dalapon was administered in distilled water once daily, by gavage, from day six through 15 of gestation (sperm positive vaginal smear = day zero). Four groups of mated females were administered zero, 500, 1,000 or 1,500 mg. of dalapon/kg./day, respectively. They were killed on gestation day 20 and fetuses were removed by cesarean section. The following tests were conducted: clinical observations, maternal body weights on days zero, six, 15 and 20, food consumption on days six, 15 and 20, observations on the number of viable fetuses, resorptions, and corpora lutea, individual fetal weights and sex, external examination, and skeletal (two-thirds) and visceral (one-third) examinations.

All rats treated with dalapon were restless for approximately ten minutes following dosing. Focal alopecia, intermittent staining of rear quarters with urine, and soft stools were observed in some treated rats. Mean weight gains and food consumption for dalapon-treated adult rats during the test interval (days zero to 20) did not differ significantly from the controls. However, mean weight gains (500 and 1,500 mg./kg./day groups) and mean daily food consumption (1,500 mg./kg./day group) were significantly less than the controls for the dalapon treatment interval (gestation period, days six to 15). Mean pup weights were significantly less than controls only in groups given 1,000 or 1,500 mg./kg./day. An increased incidence of minor variants such as retarded ossification of sternebrae was observed in fetuses of treated dams and was associated with decreased fetal weights. Spontaneous major visceral abnormalities consisted of unilateral microphthalmia in one fetus each of the 500 and 1,000 mg./kg./day groups.

In summary, no major dose-related skeletal or visceral abnormalities were observed in the fetuses of dams treated with dalapon and, other than reduced pup weights in the 1,000 and 1,500 mg./kg./day treatment levels, no untoward effects were detected.

XII. Toxicity to birds

a) Acute oral tests

The acute oral LD_{50} of dalapon sodium salt to chickens is 5,600 mg./kg. (PAYNTER *et al.* 1960).

b) Repeated daily acute oral tests

Chickens treated with ten daily repeated oral treatments using capsules containing 500 mg. dalapon sodium salt/kg./day suffered no mortality and only slight weight gain reductions over untreated chickens. No mortality or weight loss occurred under the same test conditions with dosages of 100 mg./kg./day (see Table XV) (PALMER and RADELEFF 1969).

c) Acute and subacute dietary feeding studies

Pheasants, Japanese quail, and mallards fed 5,000 ppm of dalapon sodium salt for five days suffered no mortality (HEATH 1972). Pheasant, bobwhite, and mallard populations fed 1,000 to 5,000 ppm of dalapon for various periods (ten to 169 days) sustained partial but not 100 percent mortality. Mallards appeared to be the most

Table XV. *The toxicity of dalapon sodium salt to chickens treated orally for ten days by capsule* (PALMER and RADELEFF 1969).

Dosage (mg./kg./day)	Effects
Control	No mortality, 42% wt. gain
100	No mortality, 43% wt. gain
250	No mortality, 27% wt. gain
500	No mortality, 30% wt. gain

Table XVI. *Toxic effects of dalapon in diet of various wild bird species.*

Species	Age	Amount in diet (ppm)	Test period (days)	Mortality (%)	Toxicant consumed (mg./kg./day)	Reference
Pheasant	Young	5,000	169	24	431	U.S.D.I.[a] (1964)
	Adult	5,000	10	10	250[c]	
	Adult	5,000	24	90	—	
	2–3 wk.	5,000[b]	5	0	—	HEATH *et al.* (1972)
Japanese quail	2–3 wk.	5,000[b]	5	0	—	
Bobwhite	Young	5,000	100	40	—	U.S.D.I.[a] (1963)
	Adult	5,000	100	95	610[c]	
Mallards	Young	5,000	100	24	—	
	Young	1,000	100	68	—	
	Adult	2,500	100	85	230[c]	
	Adult	1,000	100	94	120[c]	
	2–3 wk.	5,000[b]	5	0	—	HEATH *et al.* (1972)

[a] U.S.D.I. = *U.S. Department of Interior.*
[b] Dalapon sodium salt.
[c] DEWITT (1962).

sensitive species tested (see Table XVI) (*U.S. Department of Interior* 1963 and 1964).

d) Reproductive tests

The effects of dalapon when injected into eggs represents a severe test, both because of the mechanical and physiological effects of injecting chemicals through the shell and because the egg has no external excretory way to dispose of unmetabolized compounds. DUNACHIE and FLETCHER (1967 and 1969) injected from ten to 500 ppm of dalapon by weight of egg via an acetone solution into hens eggs. No teratogenic effects were seen at 500 ppm, the highest dosage tested. Hatchability of eggs was unchanged in treatments at or below 200 ppm (see Table XVII).

Table XVII. *The toxicity of dalapon injected into hens' eggs* (DUNACHIE and FLETCHER 1970).

Dalapon in egg (ppm)	Hatching season (% of control)	Teratogenic effects noted
500	70	None
400	69	None
300	63	None
200[a]	100?[a]	None
200	93	None
100	106	None
10	106	None

[a] DUNACHIE and FLETCHER (1967); 200 ppm = ten mg./egg.

Table XVIII. *Effects of dalapon sodium salt on reproduction of the mallard, pheasant and bobwhite* (DEWIFF 1962 and *U.S. Department of Interior* 1963).

Species	Dalapon sodium salt (ppm)	Eggs/female			Viable young per female[a]			% Viable young per egg[b]/laid	
		Control (A)	Test (B)	% Control/test (B/A)	Control (C)	Test (D)	% Control/test (D/C)	Control (C/A)	Test (D/B)
Mallard	2,500	27	6.8	25	5.1	2.6	51	19	38
	1,000	21	11	51	11	4.7	43	53	45
Ring-necked pheasant	5,000	32	22	68	12	6.3	55	36	29
Bobwhite	5,000	67	82	122	33	42	128	49	51

[a] Chicks per hen.
[b] Hatching success.

Dietary feeding reproduction tests conducted on adults of pheasants, bobwhites, and mallards are shown in Table XVIII. The highest dosages used in the *U.S. Department of Interior* 1963 tests (1,000 to 5,000 ppm in the diet) were checked for effects on the hatching success in terms of chicks per hen. The dosages selected were the highest of those used in Table XVI for each species and resulted in high adult mortality. Presumably part of the adult mortality occurred after the reproduction tests were finished. In any case the dosages were in the partially lethal range for adults. In spite of these high dosages, reproduction was not decreased in treated bobwhites and only partially with pheasants and mallards. Effects on reproduction were those related to number of eggs laid, a common effect at sublethal dosage levels. There was no effect on the viability and fertility (hatching success) of the three species compared to the controls.

XIII. Toxicity to fish and other aquatic animals

a) Acute toxicity to fish

The acute toxicity of dalapon sodium or magnesium salt to 12 species of fish in static water is shown in Table XIX. The LC_{50} for all of these species was greater than 100 ppm of dalapon which represents a low order of toxicity. In tests with four species of fish, 50 ppm of dalapon caused no mortality to the fry. Green sunfish eggs (embryo) were not affected by 50 ppm of dalapon (see Table XX). An acute toxicity test with dalapon in flowing water also showed a similar low order of toxicity to large-mouth bass (BOND *et al.* 1960).

Table XIX. *Toxicity of dalapon and formulations to fish.*

Species	Water temp.	Exposure (hours)	Material tested[a]	Concentration (ppm)		Mortality (%)	Reference
				Formulation[b]	Dalapon		
Rainbow trout	12°C	96	Tech.	>100(Na)	>87	50[h]	Svendson (1966)
	20°C	24	DOWPON	340F	252[c]	50	Alabaster (1969)
	20°F	48	DOWPON	210F	155[c]	50	
	50°F	24	DOWPON S	430F	318	50	Alexander (1963)
	50°F	96	DOWPON S	430F	318	50	
	50°F	96	DOWPON S	300F	222	0	
Brown trout	50°F	24	DOWPON S	345F	255	50	
	50°F	96	DOWPON S	345F	255	50	
	50°F	96	DOWPON S	300F	222	0	
	50°F	72	DOWPON S	500F	370	100	Winston (1964)
Coho salmon	68°F	24	DOWPON S	340(Na)	295	50	Bond *et al.* (1960)
	68°F	48	DOWPON S	340(Na)	295	50	
	68°F	72	DOWPON S	325(Na)	282	50	
	68°F	96	DOWPON S	310(Na)	269	50	
Bluegill	12°C	96	Tech.	>100(Na)	>87	50	Svendson (1966)
	12°C	7 Days	Tech.	—	80	0	Lawrence (1962 a)
	12°C	15–21 Days	Tech.	—	50	0	Lawrence (1962 b)
	25°C[d]	24	RADAPON	480(Na)	416	50	Surber & Pickering (1962)
	25°C[d]	48	RADAPON	440(Na)	381	50	
	25°C[d]	96	RADAPON	440(Na)	381	50	
	25°C[e]	24	RADAPON	510(Na)	442	50	
	25°C[e]	48	RADAPON	440(Na)	381	50	
	25°C[e]	96	RADAPON	290(Na)	251	50	
	75°F	24	RADAPON	—	115[g]	50	U.S.D.I. (1965)
	75°F	48	RADAPON	—	115[g]	50	
	75°F	96	RADAPON	—	105[g]	50	
Yellow perch	75°F	72	DOWPON S	300F	222	0	Winston (1964)
	75°F	72	DOWPON S	400F	296	100	
Pumpkinseed	75°F	72	DOWPON S	300F	222	0	Winston (1964)
	75°F	72	DOWPON S	400F	296	0	
	75°F	72	DOWPON S	500F	370	75	
Largemouth bass	68°F	48	DOWPON	1,000(Na)	867	0	Bond *et al.* (1960)
	68°F[f]	168	DOWPON	1,000(Na)	867	100	
	68°F[f]	168	DOWPON	250(Na)	217	0	
Channel catfish	12°C	96	Tech.	>100(Na)	>87	50	Svendson (1966)
Goldfish	12°C	96	Tech.	>100(Na)	>85	50	
	80°F	96	DOWPON	—	200	0	Hardy (1971)
	80°F	96	DOWPON M	—	200	0	
	83°F	48	Tech.	10(Na)	7	0	Lawrence (1962 a)
Fathead minnow	50°F	24	DOWPON S	480F	355	50	Alexander (1963)
	50°F	48	DOWPON S	430F	318	50	
	50°F	72	DOWPON S	410F	303	50	
	50°F	96	DOWPON S	345F	255	50	
	50°F	96	DOWPON S	200F	148	0	
	75°F	72	DOWPON	300F	222	0	Winston (1964)
	75°F	72	DOWPON	400F	296	50	
	75°F	72	DOWPON	500F	370	100	
	25°C[d]	24	RADAPON	440(Na)	381	50	Surber & Pickering (1962)
	25°C[d]	48	RADAPON	410(Na)	381	50	
	25°C[d]	96	RADAPON	390(Na)	338	50	
	25°C[e]	24	RADAPON	420(Na)	364	50	
	25°C[e]	48	RADAPON	310(Na)	303	50	
	25°C[e]	96	RADAPON	290(Na)	251	50	
Lake emerald shiner	57°F	72	DOWPON	4,000(Na)	2,960	0	Winston (1964)
Harlequin	20°C[c]	24	DOWPON	300F	222	50	Alabaster (1969)
	20°C[c]	48	DOWPON	240F	178	50	

[a] DOWPON and RADAPON are trademarks of the Dow Chemical Company.
[b] *F* = formulation containing salt of dalapon, *Na* = sodium salt of dalapon; concentrations are in terms of active ingredient.
[c] Twenty percent replacement of treated water every ten minutes.
[d] Soft.
[e] Hard.
[f] Constant flow of water.
[g] May be as sodium salt.
[h] 50 = LC_{50} value in all cases.

Table XX. *Toxicity of dalapon to fish eggs and fry*[a] (Hiltibran 1967).

Species	Exposure (days)	Dalapon (ppm)	Mortality (%)
Bluegill			
Small	12	50	0
Fry	8	50	0
Green sunfish			
Fry	8	50	0
Egg embryo	3	50	0
Lake chubsucker			
Fry	8	50	0
Smallmouth bass			
Fry	8	50	0

[a] At 22° to 25°C water temperature.

Table XXI. *The effect of dalapon on aquatic organisms in salt water* (Butler 1965).

Organism	Exposure (hours)	Temp (°C)	Dalapon (ppm)	Effect
Longnose killifish	48	29	37[a]	0% mortality
	48	20	0.87[b]	0% mortality
White mullet	48	27	37	0% mortality
Brown shrimp	48	29	0.87	40% immobilization
Blue crab	48	24	37	No effect
Eastern oyster	96	31	0.87	0% mortality or shell-growth effect
Phytoplankton	4	—	0.87	0% decrease in carbon fixation

[a] Dalapon sodium salt tested.
[b] Dalapon sodium salt tested as RADAPON herbicide formulation.

Two species of salt water fish were also tested and found to be very tolerant to dalapon (see Table XXI).

b) *Subacute toxicity to fish*

Lawrence (1962 a and b) tested dalapon at 80 ppm for seven days and at 50 ppm for 15 to 21 days without causing mortality to bluegills (see Table XV).

c) *Toxicity to aquatic organisms other than fish*

Dalapon sodium salt has been tested on a number of species of insects and other crustaceans, both in fresh water and in salt water

Table XXII. *The effect of dalapon on aquatic invertebrates.*

Organism	Exposure (hours)	Temp. (°F)	Dalapon sodium salt (ppm)	Effect	Reference
Entomostracans					
Daphnids (Cladocerans)					
Simocephalus serratus	48	60	16	EC_{50}[b]	SANDERS & COPE (1966)
Daphnia pulex	48	60	11	EC_{50}[b]	
Cyclops sp.	24	80	400 (a.e.)[a]	100 % mortality	KENAGA (1958)
	24	80	200 (a.e.)[a]	0 % mortality	
Eubranchipius sp.	24	80	400 (a.e.)[a]	100 % mortality	
	24	80	200 (a.e.)[a]	0 % mortality	
Insects					
Stoneflies, *Pteronarcys californica*, naiad	96	60	100	None apparent	SANDERS & COPE (1969)
Dragonfly, *Aeschna* sp.	48	—	1,600 (a.e.)[a]	0 % mortality	KENAGA (1958)
Snails					
Planorbis sp.	—	—	400	50 % mortality	
	—	—	200	0 % mortality	
Physa sp.	—	—	400	50 % mortality	
	—	—	200	0 % mortality	

[a] *a.e.* = acid equivalent.
[b] EC_{50} = 50 percent immobilization or mortality.

in short-exposure tests. Such tests have also been conducted on snails, oysters, and marine phytoplankton (see Tables XXI and XXII). The no-effect level of dalapon for snails, dragonflies, stoneflies, and several crustaceans is at least 200 ppm. Daphnids are more sensitive, being "immobilized" at 11 to 16 ppm. Shrimp are "immobilized" 40 percent at 0.9 ppm.

The highest concentrations of chemicals which were tested by BUTLER (1965) in estuaries where shrimp, oysters, and crabs grow are at low ppm concentrations since even ppb are not likely to occur after herbicidal use because of dilution.

XIV. Toxicity to honey bees

ANDERSON and ATKINS (1967) classified 163 pesticides into three groups based on field tests and commercial experiences for the more commonly used materials. Laboratory contact data (LD_{50} values) from ATKINS and ANDERSON (1967) were also used as follows: highly toxic <two ug./bee, moderately toxic two-to-ten μg./bee, and relatively nontoxic >ten μg./bee. Dalapon was classified in the relatively nontoxic group by these researchers. DUVAL (1969) made a similar evaluation of toxicity of pesticides to honey bees, classifying dalapon as relatively nonhazardous among the many pesticides evaluated. Laboratory tests at The Dow Chemical Company showed that honey bees immersed and wetted momentarily in dalapon sodium salt solution at a concentration of 20,000 ppm caused no mortality 24 hours after immersion.

All of the above data are in overwhelming contradiction to that of KING (1964) who stated (without data) that dalapon should not be used in areas of high bee activity because of severe contact toxicity.

XV. Toxicity to soil invertebrates (arthropods, nematodes, and earthworms)

FOX (1964) studied the effect of dalapon sodium salt on the numbers of wireworms, springtails, mites, earthworms, and millipedes in grassland soil over a two-year period after application. Since dalapon destroys grass, populations of invertebrates feeding on grass and succession plants will be affected by the kind of plants remaining. Dalapon at 40 lb./A or 20 lb./A did not affect wireworm populations, principally those of the genus *Agriotes*. A dosage of 20 lb. dalapon/A resulted in slightly increased numbers of millipedes and springtails (suborder *Arthropleona*) in the first year and increase in numbers of mites (unspecified) in the first and second year (at the 95 percent confidence limits). No significant increase or decrease in numbers of earthworms (*Lumbricidae*) occurred.

COURTNEY *et al.* (1962) found that dalapon applied at five lb./A to colonial bentgrass reduced the number of nematodes by 94 percent.

XVI. Effect on microorganisms

WORSHAM and GIDDENS (1957) found that dalapon sodium salt applied at 17, 34, and 68 lb./A caused no effect on the overall soil microbiological population (combining bacteria, actinomycetes, and fungi) except possibly to stimulate growth in some cases. Nitrates appeared to be reduced in soil one week after application of 68 lb./A dosage, however, three weeks after application, implying acclimation by the microorganisms and/or reduced concentrations due to break down of dalapon. REID (1960) studied the effect of dalapon sodium salt on nitrification rate as a measure in the changes in soil microflora at 25°C under aerated conditions. Compared to control rates of nitrification, concentrations of 25, 50, 100, and 150 lb./A gave 30, 23, 20, and two percent, respectively, 14 days after treatment and 95, 80, 77, and 48 percent, respectively, after 28 days. OTTEN *et al.* (1957) studied the effect of dalapon on nitrification by measurement of nitrate, nitrite, and ammonia-nitrogen content in soil production (*Nitrobacter* and/or *Nitrosomonas* activity). While the two dosages of 20 to 40 lb. of dalapon/A caused slightly different responses, no differences were seen in ppm of nitrates-nitrogen or ammonia-nitrogen in soil compared to control values.

BURGE (1969) found that *Bacillus* sp. acclimated to cause decomposition of dalapon in soil rapidly, but lost their ability to decompose dalapon when they were grown in nutrient broth or on nutrient agar without dalapon, showing the lack of permanent effect on such organisms. LASKOWSKI and BROADBENT (1970) studied the inhibitory effects of ten and 100 ppm of dalapon on gas (CH_4 and CO_2) and acid production by microorganisms in anaerobic clay soil. The data from these experiments indicated that the use of dalapon at rates approximating normal field application (ten to 20 lb./A) probably should have no lasting adverse effects on anaerobic systems (organisms) buffered near neutral pH.

MACKENZIE and MACRAE (1972) tested the tolerance of the nitrogen-fixing system of *Azobacter vinelandii* to dalapon sodium salt. The acetylene reduction and the growth rate of the bacterium were not affected by five and 50 ppm dalapon sodium salt in the growth media. These concentrations are equal to or in excess of those which would occur from commercial use of dalapon, even when applied directly to water.

WALSH (1972) found the sodium and magnesium salts of dalapon, the active ingredient in DOWPON M grass killer, to be four to 25 times less toxic on the basis of ten days growth at 20°C than dalapon to four species of unicellular algae in test tubes containing artificial sea water of a salinity of 30 parts per thousand (Table XXIII). Even greater differences (around 100-fold) between the acid and salts occurred in regards to decrease in oxygen production as measured by a phytosynthesis respirometer over a 90-minute exposure at 20°C. None of the organisms appeared sensitive to dalapon or its sodium or magnesium salts at concentrations which would occur in natural marine waters.

Table XIII. *Effect of dalapon and salts on microorganisms in marine waters* (WALSH 1972).

Organism	Material	Concentration[a]	
		EC_{50} (ppm)	EC_{100} (ppm)
Chlorococcum sp.	dalapon	50	100
	DOWPON M	500	1,000
Phaeodactylum tricornutum	dalapon	25	35
	DOWPON M	500	750
Dunaliella tertiolecta	dalapon	100	175
	DOWPON M	400	800
Isochrysis galbana	dalapon	20	35
	DOWPON M	500	750

[a] EC_{50} and EC_{100} = concentrations required to reduce growth of organisms 50 and 100 percent, respectively.

VENKATARAMAN and RAJYALAKSHMI (1971) found blue-green algae tolerant to dalapon at dosage rates equivalent to 200 kg./ha (179 lb./A) over a 15-day period. The species included those in the genera *Aulosira* (*fertilissima*), *Tolypothrix* (*tenuis*), *Anicystis, Anabaena,* and *Nostoc.*

XVII. Conclusions

Dalapon, as used commercially in the form of the sodium or magnesium salt, is a solid with a low vapor pressure and high water solubility. In the aqueous phase it exists principally as the ionic form. Dalapon or its salts partitions in aqueous-organic media preferentially to the aqueous system. Dalapon is hydrolyzed slowly or photolyzed principally to pyruvic acid, carbon dioxide, chloride ion, and acetaldehyde. Pyruvic acid occurs naturally as a cellular chemical involved in the respiration and energy functions of living cells of organisms not treated with dalapon. Acetaldehyde is readily oxidized to acetic acid, a common acid in nature. It appears that photodegradation represents a route of disappearance of some practical significance in the environmental fate of dalapon, but is probably of less significance than the relatively fast route of microbial degradation.

Dalapon, though not readily metabolized in animals, is a highly water-soluble compound readily excreted in the urine. All of the known degradation products from water or microorganisms are also water soluble, such as the chloride, chloropropionate, and pyruvate ions. Cows fed dalapon at 300 ppm in their diets did not exceed 2.5 ppm (averaging around one ppm in the milk during the test period. Dalapon does not partition selectively to fat tissues. Under a continuous feeding regime, dalapon is found mainly in excretory organs such as the kidney and liver, and in blood. All tissue residues in steers decrease at a half-life rate of a few days when withdrawn from feeding with dalapon. Withdrawal from a diet containing 100 ppm causes tissue residues to soon fall to less than one ppm dalapon. There is no indication of bioconcentration of dalapon in various tissues of animals or birds.

In a balanced aquatic ecosystem, there is no indication that dalapon concentrates in live fish, snails, or daphnids. Residues in dead plants and fish three to ten times those in water at five ppb are probably due to adsorption and surface/volume relationships of the organism and not due to metabolic processes. Residue build-up does occur in live plants, as expected of systemic herbicides. Dalapon does not appear to volatilize from nonflowing water in short term tests.

Dalapon is readily translocated throughout plants, concentrating in young tissues, but is very stable showing little indication of metabolism, although eventual release of ^{36}Cl and incorporation of ^{14}C in other plant components is noted in small quantities. Dalapon appears to be present essentially as dalapon (free acid) in plant tissues, and is water-extractable as such. The rate of disappearance of residues of dalapon from treated plant foliage is rapid during the first week, with about 80 percent decrease over the initial residues, and then decreases at a much slower rate over a period of several weeks or more.

Many genera and species of bacteria as well as fungi and other microorganisms in soil are capable of forming the chloride ion and pyruvic acid from dalapon or its salts. Without the presence of microorganisms, dalapon degradation in soil is slow. After an acclimation time of one to several weeks in the presence of soil microorganisms, metabolism of dalapon proceeds at an accelerated speed. Subsequently the original dosage and additional high dosages of dalapon far greater in concentration than those employed for commercial control are subject to rapid degradation. The speed of degradation of dalapon is greatly dependent on soil conditions favorable to the microorganisms and appears to be quite independent of the dosage of dalapon used. Soil conditions favorable to dalapon degradation are temperatures of 20° to 30°C, adequate soil moisture, and a pH range of 5.3 to 7.5, preferably about 6.5. Soil textures with various sandy, loam, clay, and organic matter levels appear to have a minor effect on rates of dalapon degradation.

Dalapon has essentially no effect on aerobic or anaerobic organisms in soil at dosages approximating commercial usage recommendations. A number of genera and species of algae, fungi, and bacteria tolerate high dosages of dalapon and are capable of decomposing it.

Under conditions favorable to growth of microorganisms, dalapon in soil appears to be metabolized at a faster rate than it is leached.

Microorganisms are capable of metabolizing and decomposing dalapon in water. In some instances a lag period was observed during which there was slow degradation followed by a period of rapid degradation when the microorganisms became acclimated to use of dalapon as a source of energy.

Under conditions of commercial usage on irrigation canal banks concentrations of dalapon in water are insufficient to cause damage to irrigated crops.

Mammals tolerate high concentrations of dalapon (and salts) in their diets for months and years in laboratory tests without mortality or other measurable effect. Effects caused by high concentrations are only those of weight loss and slight effects on the kidney.

Field tests with forage vegetation treated with heavy applications of dalapon sodium salt fed upon by cattle, sheep, and swine caused no ill effects with these animals.

A three-generation study showed no effect on reproduction (fertility, gestation, viability) and lactation of rats fed 3,000 ppm of dalapon sodium salt in their diets during the entire test period.

Rats fed high dosages of dalapon (up to 1,500 mg./kg.) daily during gestation suffered no significant change in the number of viable fetuses or fetus resorptions. No major dose-related skeletal or visceral abnormalities were observed in fetuses. Other than reduced pup weights in the high treatment levels, no untoward effects were noted.

Birds can tolerate high concentrations of dalapon in their diet for weeks without mortality. High dosages causing reproductive effects, as measured by egg-hatching success, are not essentially different from those causing lethal effects to breeding birds. There was no effect on hatching success with any of the concentrations causing sublethal effects or partial lethality to adults.

High dosages (residues) of dalapon injected into bird eggs do not cause teratogenic effects (>5,000 ppm) or mortality (200 ppm).

Fish can tolerate high concentrations of dalapon (usually at least 200 ppm) for one to four days and at least 50 ppm for 15 to 21 days. Eggs of fish and fry tolerate at least 50 ppm without mortality.

Most aquatic insects, crustaceans, and snails will tolerate several hundred parts per million of dalapon in acute exposures without mortality. Some crustaceans are apparently more sensitive.

Concentrations of dalapon which are likely to occur in marine waters appear to offer no significant hazard to marine organisms. Concentrations of dalapon which are likely to occur in fresh waters appear to offer no significant hazard to fish and most invertebrate organisms.

Dalapon is relatively nontoxic to honey bees. It is relatively nontoxic to soil invertebrates when used at commercial application rates to soil. Effects appear to be related to change in host-plant species resulting from the herbicide treatment, probably not much different from effects caused by soil tillage.

Residues of dalapon which may be encountered by wildlife on treated grass are representative of the highest residues occurring from commercial treatment and could be as much as several thousand parts per million temporarily. Aside from the fact that such treated grass would soon die as expected from dalapon treatments and, therefore, would be inedible for wildlife, residue concentrations from foliar treatments decline rapidly so that the highest concentrations are soon reduced to the several hundred parts-per-million range or less. Mammals and birds consuming from 100 to 1,000 ppm of dalapon in their diets even over periods of weeks, months, or years are not killed. Residues in mammal and bird tissues from such feeding remain below 100 ppm of dalapon even during treatment and soon fall below one ppm when removed from exposure to dalapon. Due to the spot or limited-area treatment practices normally used with dalapon, the mammal, bird, or fish which moves around daily in and near treated areas is not likely to contain dalapon residues in all parts of its diet.

It is concluded that dalapon poses no significant environmental safety problems, as confirmed by observations over many years of widescale commercial use. This is due to the low toxicity of dalapon to most species of plants, low toxicity to and low residue values found in nearly all exposed animal organisms, and to rapid metabolism by microorganisms.

Acknowledgments

The help given by Ralph Gantz, Milton Getzendaner, Fred Hoerger, D. D. McCollister, E. Roland Laning, Wendell Osborne, Theodore Torkelson, and Andrew J. Watson of The Dow Chemical Company in making this a better publication is much appreciated.

Summary

This review includes the physical and chemical properties of dalapon and its sodium and magnesium salts; distribution of dalapon and metabolites as residues in soil, water, plants, and animals; and toxicity of dalapon to mammals, birds, fish, other aquatic organisms, terrestrial invertebrates, honey bees, algae, bacteria, and fungi. Microorganisms under suitable growing conditions acquire the ability to rapidly metabolize dalapon in soil and water in one to two weeks. Dalapon is slowly degraded in plants and animals, in water of high or low pH, and in environments which are cold (below 60°F) or dry or otherwise inhospitable to microorganisms. Dalapon hydrolyzes in water under high temperature and light conditions. Residues of dalapon are not accumulated in tissues of animal organisms significantly above those of their diet or surrounding environment and decrease rapidly upon withdrawal from exposure to dalapon. A large amount of unchanged dalapon is excreted in the urine of mammals. Dalapon is degraded in the environment or metabolized by organisms through dechlorination, dehydrochloronation, and oxidation to chloride ion, pyruvic acid (pyruvate), alanine, carbon dioxide, and other minor or intermediate transitory compounds. Dalapon is not a hazard acutely or chronically at dosage rates used in commercial applications to mammals, birds, fish, insects, earthworms, and microorganisms. No teratogenic or significant reproductive effects have been noted in a large number of tests with birds or animals treated with dalapon. No significant environmental or commercial problems have been observed over a period of many years of widescale use. This can be accounted for by the low toxicity of dalapon to most species of plants, low toxicity to and low residue values found in nearly all exposed animal organisms, and to rapid metabolism by microorganisms. Pyruvic acid, the major metabolite of dalapon is a natural constituent of essential energy cycle systems of animal organisms.

Résumé*

Informations toxicologiques et residuelles utiles a l'evaluation environnmentale de Dalapon

Ce rapport présente les propriétés physiques et chimiques du dalapon et de ses sels de sodium et de magnésium; la distribution du dalapon et de ses métabolites sous forme de résidus dans le sol, dans l'eau, dans les plantes et dans les animaux; la toxicité du dalapon envers les mammifères, les oiseaux, les poissons, et les autres organismes aquatiques, les invertébrés terrestres, les abeilles, les algues, les bactéries et les champignons. Les microorganismes dans des conditions de croissance adéquates acquièrent la capacité de métaboliser rapidement le dalapon dans le sol et dans l'eau (dans une à deux semaines). Le dalapon se décompose lentement dans les plantes et dans les animaux, dans l'eau de pH élevé ou bas, et dans des milieux qui sont froids (au dessous de 15°C), ou secs, ou autrement inhospitaliers aux microorganismes. Le dalapon s'hydrolyse dans l'eau s'il y a une température élevée et de la lumière. Les résidus de dalapon ne s'accumulent dans les tissus des organismes animaux que peu au dessus de ceux dans leur régime ou de leur milieu environnant et ils diminuent rapidement lorsque cesse l'exposition au dalapon. Une grande quantité de dalapon intact est excrété dans l'urine des mammifères.

Le dalapon se dégrade ou se métabolise par la déchloruration, la déshydrochloruration et l'oxydation à l'ion chlorure, à l'acide pyruvique et à d'autres composés transitoires secondaires ou intermédiaires. Aux taux de dosage employés dans les applications commerciales, le dalapon n'offre pas de hasard aigu ou chronique pour les mammifères, les oiseaux, les poissons, les insectes, les vers de terre ou les microorganismes. On n'a remarqué aucun effet tératogénique ou reproductif qui soit significatif dans un grand nombre d'essais avec les oiseaux ou les animaux traités avec le dalapon. On n'a observé aucun problème significatif envers l'environnement ou aucun problème commercial pendant une période de nombreuses années d'utilisation à grande échelle. Cecil s'explique par la faible toxicité du dalapon envers la plupart des espèces de plantes, par la faible toxicité envers presque tous les organismes animaux exposés et les résidus de faible valeur qui s'y trouvet, et par le métabolisme rapide au moyen de microorganismes. L'acide pyruvique, le métabolite principal du dalapon, est un élément constituant naturel des systèmes essentiels du cycle d'énergie des organismes animaux.

---

* Traduit par The Dow Chemical Company.

Zusammenfassung*

Toxikologische und Rückstandsdaten zum Gebrauch in der Auswertung von Umweltsteinflüssen des Dalapons

Dieser Bericht behandelt die physikalischen und chemischen Eigenschaften des Dalapons und seiner Natrium und Magnesiumsalze; die Ausbreitung von Dalapon und seiner Metaboliten als Rückstände im Boden, im Wasser, in Pflanzen und in Tieren; die Toxizität von Dalapon gegenüber Säugetieren, Vögeln, Fischen und anderen Wasseroorganismen, landbewohnenden Weichtieren, Honigbienen, Algen, Bakterien und Schwämmen.

Unter geeigneten Wachstumsbedingungen erwerben Mikroorganismen die Fähigkeit Dalapon im Boden und in Wasser schnell zu metabolisieren (innerhalb von ein bis zwei Wochen). In Pflanzen und Tieren, in Wasser von hohem oder niedrigem pH und unter Bedingungen die entweder kalt (unter 16°C), trocken, oder anderweitig für Mikroorganismen ungastlich sind, wird Dalapon nur langsam abgebaut. Dalapon hydrolysiert in Wasser bei erhöhter Temperatur und unter Lichtaussetzung. In den Geweben von Tieren werden Rückstände von Dalapon nicht in Konzentrationen erheblich über denen in der Nahrung oder der Umgebung aufgespeichert, und solche Rückstände verringern sich schnell bei Zurückziehung von der Dalapon Behandlung.

Im Urin von Säugetieren wird eine beträchtliche Menge von Dalapon unverändert ausgeschieden. Dalapon wird durch Entchlorinierung, Dehydrochlorinierung und Oxydation zum Chloridion, Brenztraubensäure und anderen unbedeutenden oder zwischenstuflichen Verbindungen abgebaut. In gewerblichen Anwendungsdosen für Säugetiere, Vögel, Fische, Insekten, Regenwürmer oder Mikroorganismen stellt Dalapon keine akute oder chronische Gefahr dar. In einer grossen Zahl von Versuchen mit Vögeln oder Tieren, die mit Dalapon behandelt worden waren, wurden keine teratogenischen oder nennenswerten Auswirkungen auf das Vermehrungsverhalten beobachtet. Während eines Zeitraumes von vielen Jahren ausgedehnter Verwendung wurden keine nennenswerten Schwierigkeiten kommerzieller Art oder bezüglich der Umwelteinflüsse berichtet. Dies kann auf die niedrige Toxizität von Dalapon für die meisten Pflanzengattungen, die niedrige Toxizität für, und die niedrigen Rückstandswerte in praktisch allen behandelten Tierorganismen und auf den schnellen Abbau durch Mikroorganismen zurückgeführt werden. Brenztraubensäure, der Hauptmetabolit des Dalapons, ist ein natürlicher Bestandteil des lebenswichtigen Energiekreislaufsystems aller tierischen Organismen.

---

* Übersetzt von The Dow Chemical Company.

References

ALABASTER, J. S.: Survival of fish in 164 herbicides, insecticides, fungicides, wetting agents and miscellaneous substances. Internat. Pest. Control., pp. 29–35. Mar./Apr. (1969).

ALEXANDER, H. C.: Toxicity of Dowpon S to three species of fish. Letter to G. E. Lynn, The Dow Chemical Company, Midland, Mich., Sept. 10 (1963).

——, and T. L. BATCHELDER: Biodegradability of DOWPON® M and dalapon herbicides. Report WCL 72004. Waste Control Laboratory, Dow Chemical USA, Midland, Mich. Dec. 4 (1972).

ANDERSON, L. D., and E. L. ATKINS, JR.: Toxicity of pesticides and other agricultural chemicals to honey bees. Proc. Reg. Pest. Chem. Applicators. Riverside, Calif. Apr. 3 and 4 (1967).

ANONYMOUS: Code of Federal Regulations. 40. Protection of environment. Rev. Jan. 1, 1972. p. 425. Office of the Federal Register, National Archives and Records Service, General Services Administration, Washington, D.C. (1972).

ATKINS, E. L., JR., and L. D. ANDERSON: Toxicity of pesticides and other agricultural chemicals to honey bees. Laboratory studies. Circ. M. 16. Calif. Agr. Ext. Serv., Univ. of Calif. (1967).

BEALL, M. L., P. C. KEARNEY, and D. D. KAUFMAN: Comparative metabolism of 1-^{14}C- and 2-^{14}C-labeled dalapon by soil microorganisms. Abstr. 1964 Meeting Weed Soc. Amer., p. 12 (1964).

BLANCHARD, F. A., W. W. MUELDER, and G. N. SMITH: Synthesis of carbon-14—labeled dalapon and trial applications to soybean and corn plants. J. Agr. Food Chem. **8**, 124 (1960).

BOND, C. E., R. H. LEWIS, and J. L. FRYER: Toxicity of various herbicidal materials to fishes. Trans. 1959 Seminar: Biological problems in water pollution, pp. 96–101. The Robert A. Taft Engineering Center, Tech. Rept. W60-3. Public Health Service, U.S.D.H.E.W., Cincinnati, Ohio (1960).

BRUNS, V. F., and J. H. DAWSON: Effects of DCB, DCB-xylene mixtures, Amitrol and sodium salt of dalapon in irrigation water on corn and rutabagas. Weeds **7**, 333 (1959).

BRUST, H.: Hydrolysis of dalapon sodium salt solutions. E. C. Britton Research Laboratory, The Dow Chemical Company, Midland, Mich., Nov. 11 (1953).

BURGE, W. D.: Populations of dalapon-decomposing bacteria in soil as influenced by additions of dalapon or other carbon sources. Applied Microbiol. **17**, 545 (1969).

BUTLER, P. A.: Effects of herbicides on estuarine fauna. S. Weed Control Conf. Proc. **18**, 57 (1965).

CORBIN, F. T.: The influence of pH on the detoxification of herbicides in soil: Ph.D. thesis, N. Carolina St. Univ. Raleigh (1965).

COURTNEY, W. D., D. V. PEABODY, and H. M. AUSTENSON: Effect of herbicides on nematodes in bentgrass. Plant. Disease Reporter **42**, 256 (1962).

DAY, B. E., L. S. JORDAN, and R. C. RUSSEL: Persistence of dalapon residues in California soils. Soil Sci. **95**, 326 (1963).

DEWITT, J. B.: Toxicity of pesticides to upland birds and wildfowl. Toxicity of dalapon (sodium 2,2-dichloropropionate). Work Unit D-1.7, Bureau of Sport Fisheries and Wildlife, U.S.D.I., Patuxent, Md. (1962).

DILLON, L. S.: The science of life, pp. 42–52. New York: MacMillan (1964).

DUNACHIE, J. F., and W. W. FLETCHER: Effect of some herbicides on the hatching rate of hen's eggs. Nature **215**, 1406 (1967).

—— —— The toxicity of certain herbicides to hens eggs assessed by the egg injection technique. Ann. Applied Biol. **66**, 515 (1970).

DUVAL, C. T.: Pesticides and the honey bee. PANS **15**, 321 (1969).

EMERSON, J. L., D. J. THOMPSON, and C. G. GERBIG: Results of teratological studies in rats treated orally with 2,2-dichloro-propionic acid (dalapon) during organogenesis. Rept. HH 417, Human Health Research and Development Laboratories, The Dow Chemical Company, Zionsville, Ind., Dec. 14 (1971).

FERTIG, S. N., and M. M. SCHREIBER: Effect of dalapon ingestion on performance of dairy cattle and levels of residue in the milk. J. Agr. Food Chem. **9**, 369 (1961).

FOX, C. J. S.: The effects of five herbicides on the numbers of certain invertebrate animals in grassland soil. Can. J. Plant Sci. **44**, 405 (1964).

FOY, C. L.: Absorption, distribution and metabolism of 2,2-dichloropropionic acid in relation to phytotoxicity. II. Distribution and metabolic fate of dalapon in plants. Pl. Physiol. **36**, 698 (1961).

—— The chlorinated aliphatic acids. In P. C. Kearney and D. D. Kaufman (eds.): Degradation of herbicides, pp. 207-253. New York: Marcel Dekker (1969).

FRANK, P. A., R. J. DEMINT, and R. D. COMES: Herbicides in irrigation water following canal-bank treatment for weed control. Weed Sci. **18**, 687 (1970).

GENTRY, W. M., and C. H. VAN MIDDELEM: Determination of dalapon residues in grasses treated by DOWPON M herbicide for the control of smutgrass. Unpub. rept. GH-C 442, Agricultural Department, The Dow Chemical Company, Midland, Mich., Mar. 31 (1971).

GETZENDANER, M. E.: Determination of residues of dalapon in asparagus, apples and pears. Unpub. rept. GH-C 113, Bioproducts Department, The Dow Chemical Company, Midland, Mich., May 6 (1964).

—— Study of residues of dalapon in chicken tissues and eggs following repeated feeding. J. Agr. Food Chem. **16**, 856 (1968).

—— A study of residues in tissue of swine consuming dalapon. Unpub. rept. GH-C 338, The Dow Chemical Company, Midland, Mich. (1969 a).

—— Study of residues in tissues from steers consuming dalapon. Unpub. rept. GH-C 339, The Dow Chemical Company, Midland, Mich. (1969 b).

—— Residue study: Dalapon in grass treated with DOWPON M herbicide. Unpub. rept. GH-C 526, Agricultural Department, The Dow Chemical Company, Midland, Mich., Feb. 16 (1972).

GOLDSTEIN, H. E., and J. F. LONG: Observations on cattle, sheep and swine exposed to 2,4-D, 2,4,5-T and dalapon herbicides. Presented S. Weed Control Conf., Biloxi, Miss., Jan. 20 (1960).

HARDY,J. L.: Eflect of DOWPON M on goldfish. Letter to F. Hoerger, The Dow Chemical Company, Midland, Mich., July 29, (1971).

HEATH, R. G., J. W. SPANN, E. F. HILL, and J. F. KREITZER: Comparative dietary toxicities of pesticides to birds. Special Scientific Report—Wildlife No. 152, p. 28. Bureau of Sport Fisheries and Wildlife, U.S.D.I., Washington, D.C. (1972).

HELLING, C. S.: Pesticide mobility in soils. II. Applications of soil thin-layer chromatography. Soil Sci. Soc. Amer. Proc. **35**, 737 (1971).

HILTIBRAN, R. C.: Effects of some herbicides on fertilized fish eggs and fry. Trans. Amer. Fish Soc. **96**, 414 (1967).

HIRSCH, P., and M. ALEXANDER: Microbial decomposition of halogenated propionic and acetic acids. Can. J. Microbiol. **6**, 241 (1960).

HOERGER, F.: The metabolism of dalapon. Blood adsorption and urinary excretion patterns in dogs and human subjects. Unpub. rept., The Dow Chemical Company, Midland, Mich., July 31 (1969).

——, and E. E. KENAGA: Pesticide residues on plants: Correlation of representative data as a basis for estimation of their magnitude in the environment. Environ. Qual. and Safety, Vol. I. Stuttgart: Georg Thieme and New York: Academic Press (1972).

HOLSTUN, J. T., JR., and W. E. LOOMIS: Leaching and decomposition of 2,2-dichloropropionic acid in several Iowa soils. Weeds **4**, 205 (1956).

JENSEN, H. L.: Decomposition of choloro-substituted aliphatic acids by soil bacteria. Can. J. Microbiol. 3, 151 (1957).

KAUFMAN, D. D.: Microbial degradation of 2,2-dichloropropionic acid in five soils. Can. J. Microbiol. **10,** 843 (1964).

KEARNEY, P. C., C. I. HARRIS, D. D. KAUFMAN, and T. J. SHEETS: Behavior and fate of chlorinated aliphatic acids in soils. Adv. Pest. Control Res. **6,** 1 (1965).

——, D. D. KAUFMAN, and M. L. BEALL: Enzymatic dehalogenation of 2,2-dichlorpropionate. Biochem. Biophys., Res. Comm. **14**(1), 29 (1964).

KENAGA, E. E.: The effect of dalapon on various species of aquatic animals. Unpublished letter rept., Agricultural Chemicals Research, The Dow Chemical Company, Midland, Mich., Oct. 14 (1968).

—— Guidelines for evaluating the properties of pesticides for safe use in the wildlife environment. Down to Earth **23** (4), 11 (1968).

KING, C. C.: Effects of herbicides on honeybees. Gleanings in Bee Culture, pp. 230–233 and 250–251, Apr. issue (1964).

KUTSCHINSKI, A. H.: A laboratory method of determining the stability of dalapon sodium salt on soil and other media. Down to Earth **10** (3), 14 (1954).

——, Form and magnitude of 2,2-dichlorpropionic acid (dalapon) residues in milk. J. Agr. Food Chem. **9**, 365 (1961).

LASKOWSKI, D. A., and F. E. BROADBENT: Effect of chlorinated aliphatic acids on gas and volatile fatty acid production in anaerobic soil. Soil Sci. Soc. Amer. Proc. **34,** 72 (1970).

LAWRENCE, J. M.: Aquatic herbicide data. Agricultural Handbook No. 231, Agricultural Research Service, *U.S. Department of Agriculture,* Washington, D.C. (1962 a).

—— Aquatic herbicide data, supplement I. Agricultural Experiment Station, Auburn Univ., Auburn, Ala., pp. 1–65 (1962 b).

LEASURE, J. K.: The halogenated aliphatic acids. J. Agr. Food Chem. **12,** 40 (1964).

MACKENZIE, K. A., and I. C. MACRAE: Tolerance of the nitrogen-fixing system of *Azobacter vinelandii* for four commonly used pesticides. Antonie van Leenmenhoek **38,** 529 (1972).

MAGEE, L. A., and A. R. COLMER: Decomposition of 2,2-dichloropropionic acid by soil bacteria. Can. J. Microbiol. **5,** 255 (1959).

MILLER, P. W., and M. E. GETZENDANER: Residues of dalapon in soil treated with DOWPON® Grass Killer. Unpub. rept. GH-C 658, The Dow Chemical Company, Midland, Mich., June 5 (1973).

NELSON, N. H., and S. D. FAUST: Acidic dissociation constants of selected aquatic herbicides. Environ. Sci. Technol. 3, 1186 (1969).

NORRIS, J. M.: Acute toxicological properties of formulation M-3411 containing the sodium and magnesium salts of dalapon. Unpub. rept. BCT 2.30–45–4, Biochemical Research Laboratory, The Dow Chemical Company, Midland, Mich. Aug. 12 (1969 a).

—— DOWPON M-606. Unpub. rept. BCT 2.30–45–4, Biochemical Research Laboratory, The Dow Chemical Company, Midland, Mich. Sept. 5 (1969 b).

OTTEN, R. J., J. E. DAWSON, and M. M. SCHREIBER: The effects of several herbicides on nitrification in soil. Proc. NE Weed Control Conf. **11,** 120 (1957).

PALMER, J. S., and R. D. RADELEFF: The toxicity of some organic herbicides to cattle, sheep and chickens. ARS Production Research Report No. 106, pp. 1–3 and 23. *U.S. Department of Agriculture* (1969).

PAYNTER, O. E., T. W. TUSING, D. D. MCCOLLISTER, and V. K. ROWE: Toxicology of dalapon sodium (2,2-dichloropropionic acid, sodium salt). J. Agr. Food Chem. **8,** 47 (1960).

REDEMANN, C. T., and J. W. HAMAKER: The lactic secretion of metabolic products of ingested sodium 2,2-dichloropropionate by the dairy cow. Agricultural Research, The Dow Chemical Company, Seal Beach, Calif., July 14 (1959).

REID, J. J.: Bacterial decomposition of herbicides. Proc. NE Weed Control Conf. **14**, 19 (1960).

SANDERS, H. O., and O. B. COPE: Toxicities of several pesticides to two species of cladocerans. Trans. Amer. Fisheries Soc. **95**, 165 (1966).

—— —— The relative toxicities of several pesticides to naiads of three species of stoneflies. Limnology and Oceanography **13**, 112 (1968).

SANDFORD, H., and F. R. STOVELL: Residual phytotoxicity of amino triazole and dalapon to barley and kale. Proc. 5th Brit. Weed Control Conf., p. 321 (1960).

SHEETS, T. J. and C I. HARRIS: Herbicide residues in soils and their phytotoxicities to crops grown in rotation. Residue Reviews **11**, 119 (1965).

SMITH, G. N., and D. L. DYER: Fate of 2,2-dichloropropionic acid (dalapon) in the cotton plant. J. Agr. Food Chem. **9**, 155 (1961).

——, M. E. GETZENDANER, and A. H. KUTSCHINSKI: Determination of 2,2-dichloropropionic acid (dalapon) in sugar cane. J. Agr. Food Chem. **5**, 675 (1957).

——, Y. S. TAYLOR, and B. S. WATSON: Ecological studies on dalapon (2,2-dichloropropionic acid). Unpub. rept. NBE-16, Chemical Biology Research, The Dow Chemical Company, Midland, Mich., June 12 (1972).

SURBER, E. W., and Q. H. PICKERING: Acute toxicity of endothal, diquat, hyamine, dalapon and silvex to fish. The Progressive Fish-Culturist, pp. 164–171 (1962).

SVENDSEN, G. E.: Dalapon, sodium salt in acute toxicity studies. Mar. 11 letter to Ralph L. Gantz, The Dow Chemical Company. (1966).

SWEET, R. D., L. FEDDEMA, G. CRABTREE, and D. BAKES: Longevity of several herbicides in soils. Proc. NE Weed Control Conf. **12**, 17 (1958).

TACEY, W. J., and R. R. BELLINGER, JR.: Hydrolysis of sodium 2,2-dichloropropionate in water solution. Midland Division, The Dow Chemical Company, Midland, Mich., Oct. 16 (1958).

TANAKA, F. S., and R. G. WIEN: Hydrolysis of aqueous solutions of sodium 2,2-dichlorpropionate under self-induced alkaline conditions. Metabolism and Radiation Research Laboratory, A.R.S., *U.S. Department of Agriculture.* Fargo, N. Dak. (1972 a).

—— —— Photolysis of 2,2-dichloropropionic acid in aqueous solution. Metabolism and Radiation Research Laboratory, A.R.S., *U.S. Department of Agriculture,* Fargo, N. Dak. (1972 b).

The Dow Chemical Company: Unpublished data, several years.

THIEGS, B. J.: The stability of dalapon in soils. Down to Earth, Fall Issue (1955).

THOMPSON, D. J., C. G. GERBIG, and J. L. EMERSON: Results of tolerance study of 2,2-dichloropropionic acid (dalapon) in pregnant rats. Unpub. rept. HH-393, Human Health Research and Development Center, The Dow Chemical Company, Zionsville, Ind., Sept. 29 (1971).

TINDLE, R., D. STALLING, R. HESSELBERG, and P. BENVILLE: Comment in "Progress in sport fishery research 1969," pp. 18–19. Resource pub. 88, Bureau of Sport Fisheries and Wildlife, *U.S. Department of Interior,* Washington, D.C. (1970).

U.S. Department of Interior: Pesticide-wildlife studies: A review of fish and wildlife service investigations during 1961 and 1962. Circ. 167, Fish and Wildlife Service, Washington, D.C., June (1963).

—— Pesticide-wildlife studies, 1963. Circ. 199, Fish and Wildlife Service, Washington, D.C. (1964).

—— effects of pesticides on fish and wildlife, 1964. Circ. 226, Fish and Wildlife Service, Washington, D.C., pp. 51–77 (1965).

VENKATARAMAN, G. S., and B. RAJYALAKSHMI: Tolerance of bluegreen algae to pesticides. Current Sci. **40**, 143 (1971).

WALSH, C. E.: Effects of herbicides on photosynthesis and growth of marine unicellular algae. Hyacinth Control J. **10**, 45 (1972).

WARREN, G. F.: Rate of leaching and breakdown of several herbicides in different soils. NC Weed Control Conf. Proc., 11th Ann. Meeting, Fargo, N. Dak. (1954).

Weed Science Society of America: *Herbicide handbook*, WSSA Monograph 3, 2d ed., pp. 180–183. Geneva, N.Y.: Humphrey Press (1970).

WINSTON, A. W., JR.: Fish toxicity of dalapon and DOWPON. Letter rept. to G. E. Lynn, Waste Control Laboratory, The Dow Chemical Company, Midland, Mich., Sept. 18 (1964).

WORSHAM, A. D., and J. GIDDENS: Some effects of 2,2-dichloropropionic acid on soil microorganisms. Weeds **5**, 316 (1957).

Manuscript received October 4, 1973; accepted October 31, 1973.

Subject Index